Achieving the Sustainable Development Goals

This book draws on the expertise of faculty and colleagues at the Balsillie School of International Affairs to both locate the Sustainable Development Goals (SDGs) as a contribution to the development of global government and to examine the political-institutional and financial challenges posed by the SDGs.

The contributors are experts in global governance issues in a broad variety of fields ranging from health, food systems, social policy, migration and climate change. An introductory chapter sets out the broad context of the governance challenges involved, and how individual chapters contribute to the analysis. The book begins by focusing on individual SDGs, examining briefly the background to the particular goal and evaluating the opportunities and challenges (particularly governance challenges) in achieving the goal, as well as discussing how this goal relates to other SDGs. The book goes on to address the broader issues of achieving the set of goals overall, examining the novel financing mechanisms required for an enterprise of this nature, the trade-offs involved (particularly between the urgent climate agenda and the social/economic goals), the institutional arrangements designed to enable the achievement of the goals and offering a critical perspective on the enterprise as a whole.

Achieving the Sustainable Development Goals makes a distinctive contribution by covering a broad range of individual goals with contributions from experts on governance in the global climate, social and economic areas as well as providing assessments of the overall project – its financial feasibility, institutional requisites, and its failures to tackle certain problems at the core. This book will be of great interest to scholars and students of international affairs, development studies and sustainable development, as well as those engaged in policymaking nationally, internationally and those working in NGOs.

Simon Dalby, Professor, Balsillie School of International Affairs and Wilfrid Laurier University, Canada.

Susan Horton, Professor, School of Public Health and Health Systems, and Professor, Economics, University of Waterloo, Canada.

Rianne Mahon, Professor, Balsillie School of International Affairs and Political Science, Wilfrid Laurier University, Canada.

Diana Thomaz, Doctoral Candidate, Balsillie School of International Affairs and Wilfrid Laurier University, Canada.

Routledge Studies in Sustainable Development

This series uniquely brings together original and cutting-edge research on sustainable development. The books in this series tackle difficult and important issues in sustainable development including: values and ethics; sustainability in higher education; climate compatible development; resilience; capitalism and de-growth; sustainable urban development; gender and participation; and wellbeing.

Drawing on a wide range of disciplines, the series promotes interdisciplinary research for an international readership. The series was recommended in the *Guardian*'s suggested reads on development and the environment.

Sustainable Pathways for our Cities and Regions
Planning within Planetary Boundaries
Barbara Norman

Land Rights, Biodiversity Conservation and Justice
Rethinking Parks and People
Edited by Sharlene Mollett and Thembela Kepe

Metagovernance for Sustainability
A Framework for Implementing the Sustainable Development Goals
Louis Meuleman

Survival: One Health, One Planet, One Future
George R. Lueddeke

Poverty and Climate Change
Restoring a Global Biogeochemical Equilibrium
Fitzroy B. Beckford

Achieving the Sustainable Development Goals
Global Governance Challenges
Edited by Simon Dalby, Susan Horton and Rianne Mahon, with Diana Thomaz

Achieving the Sustainable Development Goals

Global Governance Challenges

Edited by Simon Dalby, Susan Horton and Rianne Mahon, with Diana Thomaz

LONDON AND NEW YORK

from Routledge

First published 2019
by Routledge
2 Park Square, Milton Park, Abingdon, Oxon OX14 4RN

and by Routledge
52 Vanderbilt Avenue, New York, NY 10017

Routledge is an imprint of the Taylor & Francis Group, an informa business

British Library Cataloguing-in-Publication Data
A catalogue record for this book is available from the British Library

Library of Congress Cataloging-in-Publication Data
Names: Dalby, Simon, editor. | Horton, Susan, editor. | Mahon, Rianne,
1948– editor.
Title: Achieving the sustainable development goals : global governance
challenges / edited by Simon Dalby, Susan Horton and Rianne Mahon
with Diana Thomaz.
Description: Abingdon, Oxon ; New York, NY : Routledge, 2019. |
Series: Routledge studies in sustainable development | Includes
bibliographical references and index.
Identifiers: LCCN 2018061197 (print) | LCCN 2019009657 (ebook) |
ISBN 9780429029622 (eBook) | ISBN 9780367139988 (hbk)
Subjects: LCSH: Sustainable development. | Sustainable Development
Goals.
Classification: LCC HC79.E5 (ebook) | LCC HC79.E5 A2667 2019
(print) | DDC 338.9/27--dc23
LC record available at https://lccn.loc.gov/2018061197

ISBN: 978-0-367-13998-8 (hbk)
ISBN: 978-0-429-02962-2 (ebk)

Typeset in Goudy
by Wearset Ltd, Boldon, Tyne and Wear

Contents

Illustrations

Figures

Tables

Contributors

Idowu Ajibade is an Assistant Professor in Geography at Portland State University. She focuses on urban sustainability, climate change adaptation, and societal transformation. Her current research explores how sustainability practices, adaptation projects, and future city planning exacerbate vulnerability and uneven development in the Global South. She holds a PhD. in Geography and Environmental Sustainability from Western University, Canada, and was previously a post-doctoral researcher at the Balsillie School of International Affairs.

Alison Blay-Palmer is the founding Director of the Laurier Centre for Sustainable Food Systems at the Balsillie School for International Affairs and an Associate Professor in Geography and Environmental Studies at Wilfrid Laurier University. Her research and partnerships with academics and practitioners across Canada and the United States, and internationally, enable green, healthy, just, localized food systems.

Jonathan Crush is the University Research Professor at Wilfrid Laurier University and previous CIGI Chair in Global Migration and Development, Balsillie School of International Affairs. He also holds an Extraordinary Professorship at the University of Western Cape, Cape Town. He has written and published extensively on migration and development issues in Africa including, most recently, the co-edited books: *Zimbabwe's Exodus: Crisis, Migration, Survival* (SAMP and IDRC, Cape Town, 2010), *Mean Streets: Migration, Xenophobia and Informality in South Africa* (SAMP, Ottawa, 2015) and *Diasporas, Development and Governance* (Springer, Dordrecht, 2016).

Simon Dalby is Professor at Wilfrid Laurier University, and previously CIGI Chair in the Political Economy of Climate Change, Balsillie School of International Affairs. He researches geopolitics, climate change, global security and the Anthropocene. He is author of *Security and Environmental Change* (Polity 2009) and co-editor of *Reframing Climate Change* (Routledge 2016).

Michael Egge is a doctoral candidate in Earth, Environment and Society Program at Portland State University. His research focuses on the political ecology of water in California, integrating feminist and emotional political ecology into debates on water security.

Paul Freston is Professor at Wilfrid Laurier University and *professor colaborador* in sociology at the Universidade Federal de São Carlos, Brazil. He was previously CIGI Chair in Religion and Politics in Global Context, Balsillie School of International Affairs. His books include *Evangelicals and Politics in Asia, Africa and Latin America* (New York, Cambridge University Press, 2004); and (co-edited) *The Cambridge History of Religions in Latin America* (New York, Cambridge University Press, 2016).

Alexandra R. Harrington is 2018–2019 the Fulbright Canada Research Chair in Global Governance, based at the Balsillie School of International Affairs. She holds a doctoral degree in law from McGill University Faculty of Law and is the author of the book *International Organizations and the Law*.

Jenna L. Hennebry is a Senior Research Associate at the International Migration Research Centre at the Balsillie School of International Affairs, Wilfrid Laurier University. She has carried out extensive research on labour migration and gender and is a member of the U.N. Expert Working Group on Women's Human Rights in the Global Compact and the IOM's Migration Research Leaders Syndicate.

Susan Horton is University Research Chair, School of Public Health and Health Systems, and Economics, University of Waterloo, and previously CIGI Chair in Global Health Economics, Balsillie School of International Affairs. She works on economic aspects of nutrition, public health, cancer and diagnostics in low- and middle-income countries.

Seyed Ali Hosseini is a doctoral candidate in Global Governance in the Balsillie School of International Affairs. He worked as Human Rights Officer in the UN Assistance Mission in Afghanistan (UNAMA) for over five years. He also taught and authored books about the rule of law and administrative law in Afghanistan.

Hari KC is a doctoral candidate in Global Governance at the Balsillie School of International Affairs, Waterloo, Canada. His research focuses on labour outmigration from South Asia with his PhD dissertation examining Nepal's migration policies with a concentration on women migrants in the Middle East.

Kaitlin Kish completed her PhD in the area of ecological economics in the School of Environment, Resources and Sustainability at the University of Waterloo. She is currently a postdoctoral researcher at the SSHRC-funded Economics for the Anthropocene project at McGill University and is Vice President of the Canadian Society for Ecological Economics.

Rianne Mahon is Professor at Wilfrid Laurier University and was previously the CIGI Chair in Comparative Social Policy and Global Governance, Balsillie School of International Affairs. She has published comparative work on labour market restructuring, childcare politics, and welfare regimes at local, national and global scales. In addition to co-editing numerous books she co-authored *Advanced Introduction to Social Policy* with Daniel Béland. Her current work focuses on the "gendering" of global governance through international organizations.

Stephen Quilley is Associate Professor in Social and Ecological Innovation in the School of Environment, Resources and Sustainability, University of Waterloo. Trained as a sociologist at the University of Manchester, he held positions at the Moscow School for Economic and Social Science, Keele University and University College Dublin.

John Ravenhill is Director of the Balsillie School of International Affairs and Professor of Political Science at the University of Waterloo. His work has appeared in most of the leading journals in the field of international relations. He was the 2016 recipient of the International Studies Association's IPE Distinguished Scholar Award.

Sara Rose Taylor is a Research Associate at the Conference Board of Canada and has recently completed her PhD in Global Governance at Wilfrid Laurier University. Her academic research explores governance by indicators in global social policy and feminist approaches to quantitative social science.

Vanessa Schweizer, PhD, MES is an Assistant Professor of Knowledge Integration at the University of Waterloo. Her scenario research played an influential role in the development of the Shared Socio-economic Pathways used by the Intergovernmental Panel on Climate Change, and she was a contributing author to its Fifth Assessment Report.

Diana Thomaz is a doctoral candidate at the Balsillie School of International Affairs. Her research focuses on the presence of international migrants and asylum-seekers in squatted abandoned buildings in downtown São Paulo, Brazil, and in their participation in the city's housing social movements.

Olaf Weber is a Professor and University Research Chair in Sustainable Finance at the School of Environment, Enterprise and Development (SEED), and a Senior Research Fellow at the Centre for International Governance Innovation (CIGI). His main research is in Sustainable Finance, including Sustainable Credit Risk Management, Sustainable Investing and Banking, Social Banking, Impact Investing, Financial Sector Sustainability Regulations, and Codes of Conduct.

Alan Whiteside was the founding director of the Health Economics and HIV and AIDS Research Division of the University of KwaZulu-Natal where he worked from 1983 to 2013. He was awarded the Order of the British Empire in recognition of his work in AIDS and health in 2015. He is Professor at Wilfrid Laurier University and previously was the CIGI Chair in Health Policy, Balsillie School of International Affairs.

Laine Young is a PhD Candidate (ABD) in Geography at Wilfrid Laurier University and is affiliated with the Laurier Centre for Sustainable Food Systems. She integrates a feminist lens to her dissertation research in Quito, Ecuador to understand the inequity and unequal power relations present in experiences of urban agriculture.

Foreword

John Ravenhill

The Sustainable Development Goals (SDGs), adopted by the UN General Assembly in September 2015, built on decades of multilateral co-operation in the promotion of economic development. They were the culmination of three years of intensive high-level negotiations to define successors to the Millennium Development Goals. They built on unprecedented consultations with civil society actors and members of the public from almost all the UN member states. No one could accuse the global community of lacking in ambition in setting out 17 broad goals to be addressed through meeting no less than 169 individual targets by 2030. For the first time, the global community attempted to produce an action agenda that provided a coherent integration of diverse development issues. In doing so, it elevated sustainability to the forefront of the international agenda, a recognition that in the era of the Anthropocene a lack of progress on global environmental issues threatens to undermine progress in all other areas of human development.

When the Balsillie School of International Affairs (BSIA) was contemplating a publication to celebrate its tenth anniversary, a project on the SDGs was an obvious candidate to showcase the research of its faculty and students. The School was established in 2007 thanks to generous gifts from Mr. Jim Balsillie, co-founder of Research in Motion (later renamed after its best-known product, Blackberry). The School is a partnership between two universities – the University of Waterloo, and Wilfrid Laurier University – and a public policy think tank, the Centre for International Governance Innovation (CIGI), which gives it a unique structure. A separately constituted not-for-profit corporation, the School was intended to be an academic complement to the think tank that Jim Balsillie had also founded, providing graduate training and conducting high quality research. The Balsillie funding supports students in the graduate programs that the School hosts, and it provided academic leadership to the new School through a dozen senior appointments to research chairs (the CIGI Chairs Program, seven of whom have contributed to this publication).

The collaborating institutions bring to BSIA different but complementary strengths, roles, and responsibilities. The two universities employ BSIA faculty and offer BSIA's academic programs while CIGI, as a think tank, uses its in-house expertise and its worldwide network of practitioners to help inform and guide BSIA's outreach and collaborative research. The unique integration of the collaborating institutions' approaches and cultures gives BSIA an unmatched ability to promote vigorous engagement across boundaries of discipline and practice, to connect today's experts with tomorrow's leaders in critical debate and analysis, and to achieve – in all of its work – the highest standards of excellence.

In the decade since its foundation, the BSIA has quickly established itself as a leading international institution for graduate training and research in global governance and international public policy. With the completion in 2011 of the construction of the award-winning CIGI campus, where the BSIA is housed, the School's development accelerated. Enrolments in the multi-disciplinary graduate programs that the School hosts for its two university partners have grown significantly, the School currently admitting around 50 Master's students and 12 doctoral students each year. The School has established a number of exchange arrangements with leading universities in Asia, Australia, and Europe. The relationship established with Global Affairs Canada, through which our students write policy papers for the department's Foreign Policy Bureau, provides a unique opportunity for graduate students to participate in the policy-making process.

More than 60 faculty from its 2 partner universities are involved in the School's graduate programs: a further 25 are affiliated with its wide-ranging research activities. To better define its research agenda, the School has established seven research clusters. These co-ordinate events and faculty and student research in the areas of conflict and security; environment and resources; global political economy; indigenous peoples decolonization and the globe; migration, mobilities and social politics; multilateral institutions; and science and health policy. The clusters have helped to promote collaboration across the three partners and to build links with areas in the universities not traditionally associated with the School.

The School, which now hosts more than 120 events each year, has created a lively research climate and community. In support of its research strategy, the School has also hosted a number of major research centres and international projects: the Canadian Network for Research on Terrorism, Security and Society (TSAS), the International Migration Research Centre (IMRC), the Centre for Sustainable Food Systems, and the Secretariat of the Academic Council on the United Nations System; the Armageddon Letters; the Canadian Network for Defence and Security Analysis; and the Hungry Cities Partnership.[1] One of the things in which the School takes particular pride is involving graduate students in its research projects. A notable feature of the current book is that five of its authors are current or past PhD students or postdoctoral scholars at the School.

The SDGs mirror the ambitious mission statement of the School: "to develop new solutions to humanity's critical problems, improve how the world is governed now and in the future, and contribute to enhancing the quality of people's lives". The SDGs map neatly onto the School's research clusters.

The contributions to this book reflect some of the work from the clusters. It covers many of the highest profile SDGs, notably climate action, quality education, gender equality, zero hunger, and good health and well-being. And, the inter-relationship between many of the SDGs makes it impossible for the authors to discuss these goals meaningfully without touching on others such as reduced inequalities, sustainable cities and communities, and peace, justice and strong institutions. A book on the SDGs would be incomplete without an analysis of the broader issues that will influence their implementation, most fundamentally the challenges involved in raising sufficient funds to finance the ambitious agenda. These challenges are detailed in the second part of the book.

Inevitably, compromises were required to bring a set of complex negotiations to a successful conclusion. As our authors point out, even with its lengthy list of goals and specific targets *The 2030 Agenda for Sustainable Development* contains significant omissions. In addition, although the framers of the agreement displayed an unprecedented sensitivity to the inter-relationships between the multiple targets, important questions remained unaddressed as to whether some goals can be met without undermining progress towards the realization of others. Finally, there were inevitably questions of whether the voluntary process of compliance was an effective one. These pitted critics who favoured a more radical transformative agenda against pragmatists who believed that the agreement was the best that could be achieved in the circumstances, that it would concentrate international attention on the development agenda, and that the soft law approach of monitoring and information sharing was the only practical means of attempting to induce compliance.

The year 2015 may well be the high spot for internationalism in the first quarter of the twenty-first century. Not only was agreement reached on the SDGs but three months later the international community also signed the Paris Agreement on Climate Change. This volume celebrates these achievements but, in identifying unresolved issues, points the way forward for further research and action.

The School is particularly grateful to Alan Whiteside for having suggested this project and to Simon Dalby, Susan Horton, Rianne Mahon, and Diana Thomaz for all their work in bringing it to a successful conclusion.

Note

1 To find out more please visit www.balsillieschool.ca/research/. Accessed 12 December 2018.

1 Global governance challenges in achieving the Sustainable Development Goals

Introduction

Simon Dalby, Susan Horton and Rianne Mahon

Introduction

The Sustainable Development Goals (SDGs) were officially adopted in September 2015 by the United Nations General Assembly. With 17 goals and 169 targets, they are, by any standard, an ambitious list of aspirational statements. This attempt at agenda setting for the globe is not one that sets out to maintain the status quo, to manage existing practices and procedures better, or to co-ordinate incremental changes. As the title of the official United Nations agenda document signals, it aims at nothing less than "Transforming our World: The 2030 Agenda for Sustainable Development".

The theme of transformation is important in what follows because the agenda laid out by the goals clearly requires fundamental changes to numerous societal practices and rapid innovation across diverse societies. It is not, as much of the United Nations Security Council activity frequently is, simply responding to events to mitigate suffering or to attenuate conflict. It is about much more than traditional themes of international relations, the rivalries of great powers, the dangers of conflict, the co-ordination difficulties of international trade, or the protection of human rights. This is an altogether more ambitious set of aspirations and one that, because it tackles so many facets of human life, is more properly considered a matter of global governance rather than the more narrowly focused "high politics" of diplomacy, competition, and rivalry in traditional international relations. As such this volume addresses these questions explicitly in terms of governance broadly construed.

The Goals require co-ordination and administration across sectors and societies presenting those charged with its implementation unprecedented governance tasks over a 15-year period. It is heady stuff full of universal ambition, but its implementation will depend on states in very different contexts, tackling these issues in their own particular ways. As the Paris Agreement on Climate Change, finalized three months later, emphasized, universal aspirations are to be accomplished in particular ways in the specific situations that applied to particular states (Falkner 2016). As such, given the lack of overarching authority, enforcement mechanisms, or legal arrangements in the Goals program, governance is more about legitimacy, accountability, the mobilization of technical capabilities, and popular support than it is about traditional modes of state command and control.

While the role of states is obviously a key part of the process, the larger co-ordination, monitoring, reporting, and implementing functions will, the Goals' authors hope, incorporate more actors into a revitalized "Global Partnership for Sustainable Development". Led by activists and functionaries in numerous insti-tutions – the modern missionaries (Freston, Chapter 10) – operating at multiple scales, such partnerships are to advocate and innovate to transform societies through sustainable and inclusive economic growth. Given the multitudinous technical processes and academic disciplines involved, the administrative tasks are enormously complex. Nor is there reason to believe that the Goals are necessarily compatible with one another. The governance challenges include continuing political contestation over priorities and identification of appro-priate indicators. Moreover, funding new initiatives is, in light of the history of inadequate supply of foreign aid and investment from developed states (see Horton, Chapter 13), a problem for many of the Goals.

Nonetheless, for all the difficulties with the formulation and the implemen-tation of the Goals, they represent a major milestone in the emergence of what is now properly called global governance (Zürn 2018). Putting them in this context requires first looking back over the last couple of decades and their emergence from the prior programs of development. Some of the difficulties in previous arrangements were the stimulus for the process which led to the SDGs. This introduction also offers a broad sketch of the difficulties of implementation of the SDGs to identify some of the dilemmas addressed in greater detail in later chapters. Given the huge agenda not all aspects of the SDGs can be investi-gated within the covers of one volume. What this book does offer is a series of chapters focused on important aspects of the SDGs. It does so, reflecting the inter-disciplinary ethos of the Balsillie School, with its emphasis on tackling global governance issues from a variety of intellectual viewpoints.

Beyond the MDGs

The SDGs follow the original Millennium Development Goals (MDGs) in the choice of governance technique – agenda-setting through selection of a set of common goals and targets, supplemented by indicators to monitor progress. Yet the SDGs go well beyond the MDGs in important respects. While the MDGs focused on the Global South, the SDGs are universal. As Razavi (2016, 28) notes,

> The 2030 Agenda's universal application means that it is not merely "our agenda" for "them".... Rather it is a global template for a world that is increasingly integrated through flows of finance and people, in which poverty, deprivation, inequality … and unsustainable patterns of produc-tion and consumption, are as much a concern in the rich advanced economies as they are in the developing world.

More specifically, the MDGs aimed to halve the number of people living in extreme poverty, defined as a daily income less than US$1.25 a day. This absolute measure offered a narrow definition that fails to take into account other important aspects of well-being (Deacon 2014, 27), while also ignoring the very real, if "relative", poverty experienced in wealthier parts of the world. Although SDG 1 retains this absolute definition of poverty, its second target introduces a relative definition ("to reduce at least by half the proportion of men, women and children ... living in poverty in all its dimensions *according to national definitions*" [UNGA 2015, 15, emphasis added]). SDG 1 also reiterates the commitment to implementing the global social protection floor undertaken by UN agencies, member states, the development banks, and key international NGOs in 2013.[1] More importantly, the SDGs go beyond a focus on poverty to include a stand-alone goal on the reduction of inequality within and among countries: in other words, the SDGs promise to tackle the unequal distribution of resources globally and at the national scale. Whereas the MDGs effectively sidelined the Education for All (EFA) agenda that focused on education quality, early childhood education, secondary education, adult literacy, and attention to marginalised and vulnerable populations (Fukuda-Parr, Yamin, and Greenstein 2014, 110), SDG 4 embraced the EFA coalition's position (Unterhalter 2019). More broadly, whereas the MDGs focused on a limited set of social priorities, the SDGs aim to encompass a richer definition of social goals and to simultaneously address economic, social, and environmental dimensions of development on a global scale.

The adoption of such a potentially transformative global agenda was not, however, a foregone conclusion. The initial vision for post-2015 was more along the lines of an MDG + 1 (Fukuda-Parr and Hegstad 2018). The 2010 High-Level Plenary of the United Nations General Assembly (UNGA) on the MDGs had requested the Secretary General to initiate thinking on a post-2015 development agenda. This launched a process involving 90 national consultations, 11 thematic consultations, an online platform (The World We Want 2015) and MYWorld, a survey that included people from over 190 countries (Kamau, Chasek, and O'Connor 2018, 82–83). The Secretary General also established a UN System Task Team and a High-Level Panel on Post-2015, co-chaired by the then-prime minster of the UK, and the presidents of Indonesia and Liberia. While the Task Team's report, "Realizing the Future We Want for All" (UNDESA 2012), highlighted a number of the MDG's lacunae,[2] the High-Level Panel's report, *A New Global Partnership: Eradicate Poverty and Transform Economies Through Sustainable Development* (2013), favoured an approach that built on the MDGs (Fukuda-Parr and McNeill 2019).

While the post-2015 process was unfolding, preparations were being made for the UN Conference on the Environment (Rio + 20). In light of the failure of the Copenhagen Climate Conference, Colombian Paula Caballero Gómez[3] persuaded the organizers to opt for "an open, inclusive and transparent" process (Kamau, Chasek, and O'Connor 2018, 40). The Open Working Group (OWG),[4] established by the UNGA in 2013, was charged with developing a sustainable development agenda through discussions with member states and representatives of the nine major groups.[5] It was only at the UNGA Special Event, September 2013, that the two processes were merged and the OWG's approach for negotiating the goals was adopted. This meant that the SDGs would be developed through dialogue that included a range of non-state actors, in marked contrast to the MDGs, which emerged from a narrow technocratic process that reflected donor country priorities (Fukuda-Parr and Hulme 2011). The choice of this channel also favoured the adoption of the principle of universality.[6]

Forces favouring an MDG + 1 agenda continued to try to influence the outcome. For instance, the African Group expressed concern that the SDGs would divert resources from the MDGs, which had still to be met especially in the least developed countries. The UN Secretary-General and the president of the General Assembly also feared that the MDGs could be submerged in the SDGs (Kamau, Chasek, and O'Connor 2018, 98). Among others, the Australia, Dutch, and UK troika favoured continuing the MDGs' focus on the eradication of extreme poverty. In addition,

> some developed countries were concerned about the SDGs being a "universal" agenda. They were not comfortable with the United Nations prescribing what they had to accomplish, and they much preferred the existing system, where they engaged in development activities in developing countries and were not held accountable by the United Nations for sustainable development at home.
>
> (Kamau, Chasek, and O'Connor 2018, 111)

Throughout the discussions, one of the important issues of contention was whether and how to incorporate inequality. In Latin America, the "pink tide" of leftist governments had made tackling inequality a key objective, while the 2008 financial crisis and its aftermath, including the protests of groups like Occupy Wall Street, helped shed light on deepening inequality in the North. The Organisation for Economic Co-operation and Development's (OECD) two reports – *Growing Unequal: Income Distribution and Poverty in OECD Countries* (2008) and *Divided We Stand: Why Inequality Keeps Rising* (2011) – provided documentary evidence supporting the protesters' claims as did Piketty's *Capital in the Twenty-first Century* (2014). Addressing the root causes of inequality was also seen as a central task coming out of the Rio + 20 meetings. What drove the point home for the OWG, however, was Joseph Stiglitz's keynote address at the 2014 round table, which highlighted the fact that in the US 95 per cent of the income gains since 2009 had accrued to the richest 1 per cent (Kamau, Chasek, and O'Connor 2018, 94).

A standalone goal on inequality within and between countries remained contentious until the end, with China and the G77 in favour and many OECD countries opposed (Fukuda-Parr 2018).[7] While the post-2015 consultations had underlined the importance of tackling vertical (the concentration of wealth at the top) and horizontal (exclusion of the poor and vulnerable from developing their capabilities) inequality, the High-Level Panel's final report focused on the latter by embracing the concept "leave no one behind".[8] In addition, the World Bank succeeded in making its definition of inequality – focused on the absolute income growth of the bottom 40 per cent while ignoring the increased concentration of wealth in the top 1 per cent – the target for SDG 10.1. This bias is reproduced in the choice of indicators for Goal 10, none of which capture trends in the distribution of income within and between countries (Fukuda-Parr 2019).

Inequality may have been the Achilles' heel of the MDGs. Few can argue against a goal of reducing poverty, and reducing poverty is a key aim of some international organizations, for example "(t)he overarching mission of the World Bank Group is a world free of poverty" (World Bank 2013, 9). Despite criticism, and the rather arcane way that the $1.25-a-day poverty yardstick was developed, the MDGs were accompanied by other dimensions of poverty reduction, exemplified by the improvements in primary school enrolment, and reductions in child and maternal mortality rates as well as stunting (United Nations 2015). At the same time, it proved extremely difficult to reach the "Bottom Billion" (Collier 2007) who are in conflict situations, fragile states, or highly marginalized in more stable countries.

Reducing inequality is tricky enough within countries due to opposing interests, and international governance aimed at doing so internationally is extremely weak. Overall, global inequality among individuals decreased slightly (or at least did not increase) over the early period of the MDGs according to Lakner and Milanovic's (2013) painstaking analysis of data for 1988 to 2008. This was largely due to substantial growth in both India and China which were categorized as low-income countries at the start of this period. However, this growth was accompanied by (possibly even at the expense of) virtual stagnation of incomes of those at the 85th percentile in the global distribution of income, largely blue-collar workers in the high-income countries.

A growing body of research (particularly by the World Inequality Lab participants, see, for example, Alvaredo *et al.* [2018]) suggests that income distribution within the high-income countries has worsened over the past two decades, particularly at the very top. Some of this is affected by a lack of political will domestically, by permitting devices such as trusts and by not utilizing instruments such as inheritance taxes (Piketty 2014). Global competition has also led to a "race to the bottom" in terms of declining taxes on corporations and reduction in income tax rates on high earners. This is also exacerbated by technological developments which have allowed international corporations to utilize loopholes in international policy co-ordination on taxation and financial regulations. The FANG (Facebook, Amazon, Netflix, and Google) and other similar beneficiaries of a globalized, tech-intensive system utilize perfectly legal disjoints between national policies to, for example, headquarter their global international property rights in Ireland, and by means of devices such as the "double Irish" and "Dutch sandwich" to move profits to Bermuda and similar jurisdictions (Kahn 2018). Initiatives are underway such as BEPS (Base Erosion and Profit Sharing) by the European Union (OECD 2018), and tightening regulations on banking havens to disclose previously secret information on accounts, but much less quickly than international corporations can transfer funds. The end result has been initiatives by private individuals to expose what is happening by creating leaks such as the Panama Papers (ICIJ 2017b) and the Paradise Papers (ICIJ 2017a). Given the vested interests involved, dealing with these forces favouring inequality will be an extremely thorny issue for the SDGs.

There were other contentious issues. The concerns of some important groups, notably migrants (see Crush, Chapter 6), and indigenous peoples received scant attention, while the states of the North managed to keep other issues – such as tax evasion and regional and bilateral trade agreements – off the agenda. Several African and Middle Eastern countries opposed inclusion of LGBTQ and sexual and reproductive rights, while the Nordic countries championed the latter.[9] The compromise is reflected in the wording of Target 5.5:

> ensure universal access to sexual and reproductive health and reproductive rights *as agreed in accordance with the Programme of Action of the International Conference on Population and Development and the Beijing Platform for Action and the outcome documents of their review conferences.*
>
> (UNGA 2015, 18, emphasis in original)

Unterhalter (2019) documents the efforts of the EFA movement, advocating a comprehensive approach to education against, inter alia, Jeffrey Sachs' Sustainable Development Solutions Network, which focused on learning outcomes for children and youth. As she notes, while Goal 4 ostensibly embraces quality education and the promotion of lifelong learning opportunities for all, the indicators chosen undermine these progressive objectives. Blay-Palmer and Young (Chapter 2) note the disjuncture between the ambitions behind Goal 2's targets and the indicators chosen.[10] More broadly, building on the growing body of literature on "governance by numbers",[11] Fukuda-Parr and McNeill (forthcoming) are critical of the ways in which quantification too often distorts, or even perverts, goal achievement.

There are also concerns about the SDGs' reliance on the private sector, which is not surprising given the latter's access to numerous avenues of influencing the outcome. Thus, Razavi (2016, 28) notes that:

> The corporate sector … has been in a far more privileged position to influence the agenda, not only through its own Major Group (Business and Industry) but also through key bodies and channels such as the Secretary-General's High-Level Panel of Eminent Persons on the Post-2015 Agenda and the Global Compact, while also having a voice through the intergovernmental process.

The UN's turn to the private sector can be traced back to the 1990s (Bidegain Ponte and Rodríguez Enríquez 2016, 90–91). Horton and Weber (Chapter 13 and Chapter 14), however, both argue that private sector involvement is key to financing the SDGs.

SDG implementation

Implementation of the goals is to a very substantial extent a matter of governance and co-ordination among numerous agencies, but in a context where national governments are key to success. This set of goals is a relatively novel effort in global governance, and as such represents a departure from top-down regulation or market-based approaches. In contrast, "the SDGs promise a novel type of governance that makes use of non-legally binding, global goals set by the UN member states" (Biermann, Kanie, and Kim 2017). Four key points follow from this innovation. First, as the goals are not legally binding, there is no mechanism to enforce compliance. Second, although the High Level Political Forum on Sustainable Development is a novel arrangement replacing the earlier Commission on Sustainable Development, the institutional arrangements for implementation and oversight at the intergovernmental level are weak because the implementation process of the SDGs is conducted by voluntary national initiatives. The High-Level Forum convenes periodic United Nations meetings to monitor progress, and it can highlight shortcomings and failure to reach promised targets, but the key to successful implementation are bottom-up initiatives with partnerships between stakeholders within countries and with international organizations. Third, the SDGs were agreed to by a preliminary goal-setting process with input from numerous governments rather than being a top-down arrangement driven by the UN secretariat. Finally, the goals allow much flexibility on the part of individual states to prioritize which goals they pursue and how. The Paris Agreement, finalized a few months after the formal adoption of the SDGs, follows a similar model in hopes of facilitating action that is appropriate for particular contexts (Falkner 2016).

In practical terms measuring progress is key to successful implementation. This in turn requires agreement on how to collect and interpret statistical information. Inevitably it will also require new forms of research and investigations about how local performance of the goals has global consequences. Which indicators matter most and where is not a simple matter; nor is it a simple matter to co-ordinate implementation across goals or across multiple stakeholders, including civil society organizations, corporations, and governments. Given the number of goals and the scope of their ambition, implementation is bound to pose a series of major challenges. Goal 17 is explicitly designed to build the institutional partnerships needed to facilitate implementation such as the Sustainable Development Solutions Network with offices in Paris, New Delhi, and New York, which aims to mobilize expertise from the policy and scholarly networks to provide guides, policy briefs, and research materials to support SDG implementation.

Economic growth that increases wealth, but at the long-term cost of environmental destruction, eventually undercuts the gains, especially for very poor people in vulnerable locations. Thus, integrated planning of economy, society, and environment together constitutes the gold standard for the SDGs, but how to transcend the traditional policy silos poses a key question for all stakeholders.

Clearly co-ordination is important but innovation in terms of what kinds of knowledge are produced, by whom, and in what formats will be necessary. The challenge of SDG implementation includes work in the academy too, which will require inter-disciplinary research to think about cross-cutting issues between the goals.

The High Level Political Forum has a central role in overseeing the follow-up but:

> the High Level Political Forum is ... based on voluntary country-level reviews without any universal mechanism to assess each country's contribution to the global realisation of these goals, nor to review and monitor multilateral agencies, the Bretton Woods institutions and any corporation or "partnership" wanting to use the UN name, logo or flag.
>
> (Esquivel 2016, 13)

The dangers of the SDGs being used by numerous agencies and institutions to further their own interests cannot be ruled out nor is it clear that the High Level Political Forum will be effective in policing national oversight of this problem. A different option, built along the lines of the Human Rights Council's Universal Periodic Review mechanism, might have been chosen (Razavi 2016, 38–39). Such an option would have entailed a more robust peer-review process and could have included shadow reports from civil society.

Nonetheless, reporting on the SDGs is happening and while much of it depends on national statistical systems devised for other purposes, there are efforts being made to supplement these. The Sustainable Development Support Network, with technical support from the German development and education foundation Bertelsmann Stiftung, has been compiling a series of national measures in "dashboards" that monitor progress each year (Sachs *et al.* 2018). These comprehensive reports include national figures, and "league tables" of which state is performing best according to key indicators. The initial dashboard indicators focus on the G-20 countries, understood to be the large key states whose implementation of the goals will be key to their overall success.

As of 2018 no state was on track to meet all the goals. Hence implementation is sluggish, at least so far, given the lack of leadership by the major states. Scandinavian states are furthest along in implementation but even here the "dashboard" analysis suggests that work on Goal 12 on sustainable consumption and 13 on climate change still need attention. Developed countries seem to be paying little attention to the world's oceans, the source of essential protein for people in many places in the world, and they have done little to deal with biodiversity issues on land (Goal 15). Disaggregated data suggest that inequality in many of the G20 countries means that despite high average figures on economic performance not everyone in these states benefit. Conflict areas not surprisingly are having great difficulty with many goals, not least 1 (poverty) and 2 (hunger).

Given the sheer number of targets, contradictions between the goals are likely. While energy is clearly needed as part of many development programs, if it is supplied by coal power-generating stations, then climate change will be exacerbated. If limited resources in the least developed states are focused on only a couple of goals, it is likely that others will be neglected. International donors who prioritize particular issues may skew the allocation of funds and expertise in ways that are not necessarily in line with national priorities. Moreover, there is no necessary connection between priorities in national capitals and those of rural peripheries. In rural areas, women's equality may be blocked by patriarchal systems of land holding. These obstacles may, in turn, stymie attempts to improve nutrition. Hence the emphasis in the SDG implementation documents on dialogue among stakeholders and the provision of expert advice on moving the agenda ahead. If business as usual is no longer acceptable, given the environmental problems that it generates, then innovations in social life and administrative practices by government are unavoidable.

Promoting social change and removing obstacles to implementation of the SDGs thus cannot avoid the practicalities of high politics. Rose Taylor and Mahon (Chapter 4) talk about the SDGs as a space of contestation that is less than ideal but has potential to initiate changes at various scales. From the small scale of local communities through national governments and at the largest scales of geopolitical rivalries, effective implementation requires that actors accept the need for the goals and are willing to actively work towards their implementation. The unanimity of the adoption of these wide-ranging targets is noteworthy, and as such opens up the space for contestation. However, inequality, environmental destruction, and widespread poverty remain stubborn problems, even if some substantial progress, such as the reduction of childhood mortality, was accomplished by the MDGs. Resistance to social change is widespread, and the globalization backlash expressed by right wing populist movements makes implementation all the more difficult, because of direct policy opposition to global initiatives as well as the lack of financial support.

Many of the aspirations of the SDGs require international co-operation and leadership from the major powers to provide funds and co-ordination. Donald Trump's election as American president has produced an administration much more concerned with national priorities and the reduction of government activity, the deregulation of industries and abandonment of international arrangements, both formal and informal, not to mention the gutting of numerous initiatives to protect health and the environment. While in the long run the Trump administration may not derail global governance, it can certainly curtail the implementation of the SDGs, and given the urgency of tackling climate change, the delay in acting on the Paris Agreement makes everything more difficult. Crucially, as Selby (2018) argues, the preoccupation within the Trump administration about competition with China and fears of declining hegemony identifies great power rivalry, not climate change, as the most important matter of politics.

Part of the SDGs' appeal is precisely the development of a common agenda and hence a process that ought to reduce at least some of those rivalries. How well the SDGs are implemented in the coming decade will depend in part on how well they work to deliver on the initial promise of the United Nations for a peaceful and secure planet. Certainly, the Trump administration's focus on "America first" downplays the necessity of acting collectively to deal with common problems. The global agenda sketched out in the SDGs gets short shrift from an administration that insists that the context for policy is one of competing national states not a planet facing the need to simultaneously deal with poverty, illness, and inequalities across numerous societies. Despite China's rising influence, it is far from clear that its Belt and Road initiative in particular operates as a global leadership effort on the SDGs or anything else, rather than as a means for expanding its trading arrangements and political influence. If the Earth is one, the world still is not. Multipolarity and regional rivalries remain a major obstacle to the goals' implementation.

The SDGs and governance: coverage of this volume

The individual chapters in this book offer diverse perspectives on the global governance challenges to achieving the SDG. Given the interrelated nature of the SDGs, most of the chapters, even if primarily focused on a particular SDG, speak to several others. Five of the seven chapters in the first half of the book focus on a specific SDG – SDG 1 on food; SDG 2 on health, SDG 5 on gender, SDG 11 on cities, and SDG 13 on climate action. The other two chapters in this section reflect on a "notable silence" in the SDGs, namely migration. The second group of seven chapters analyses the SDGs as a group. Six of these examine either SDG 16 (Peace and Institutions) or SDG 17 (Partnerships), and challenges to the overall achievement of the SDGs. Chapter 11 rejects the SDG's claim to promoting sustainability and offers an alternative set of goals. Regrettably, given the scale of issues involved, it was not possible to cover all the SDGs. Five of the SDGs (SDG 4 on education, SDG 6 on water and sanitation, SDG 11 on industry, SDG 13 on life in water, and SDG 14 life on land) receive scant attention. Two of the arguably most important – poverty (SDG 1) and inequality (SDG 10) – are not covered by individual chapters but are a thread that runs throughout the whole volume.

The chapters come from varying theoretical perspectives and disciplinary backgrounds, with a fairly even split between a critical social science or political economy approach (Blay-Palmer and Young, KC and Hennebry, Ajibade and Egge, Dalby, Freston, Hosseini); and a more pragmatic or technocratic focus (Whiteside, Crush, Horton, Weber, Schweizer, Harrington). Coming from a feminist standpoint shared with Young and Blay-Palmer and KC and Hennebry, Rose Taylor, and Mahon offer a somewhat hybrid approach, which discusses the value of working within the structures of the SDGs while stressing the import-ance of contestation throughout their implementation. Quilley and Kish fairly comprehensively reject the vision of the SDGs and visualize a very different (a high-tech traditionalist, communitarian) society.

Although all authors agree that the SDGs agenda needs to be a trans-formative one, not all agree on the necessary scale of this transformation.

Blay-Palmer and Young (Chapter 2) see a radical shift in agriculture as required, advocating an agroecology approach that incorporates traditional knowledge and emphasizes women's role, and the role of community. Their view can perhaps be encapsulated by their comment that "it is important not to subsume agro-ecology into the industrial food system but for it to activate transformative change as a social movement" (p. 31).

KC and Hennebry (Chapter 5, pp. 71–85: one of two chapters on migration) argue that ultimately "(t)he SDGs do not address the more structural and pervasive nature of gender discrimination and structural violence that is deeply embedded in the existing governance paradigm" and that more profound changes are required. They use a case study of Nepal, one of the countries with the highest share of national income coming from remittances, to highlight the implications for the welfare of women migrants of ambivalent national policy in the sending country combined with disregard in the receiving countries.

Noting that SDG 13 (climate) is the only goal whose agenda is specified as "urgent", Dalby (Chapter 8, pp. 117–131) argues that a game-changing agenda is required: "in the current circumstances of the Anthropocene, developing sustainability requires a drastic cut in the use of combustion in the affluent metropoles of the global economy". Desperate times call for desperate measures. Dalby (Chapter 8, p. 126) argues that national governments will not be able to take actions on the required scale, and that international action is required: "the SDGs 13 on climate and 16 on Justice and Institution Building suggest the need to think through responses to climate change in more dramatic ways than much of the conventional discussion has so far considered".

Hosseini (Chapter 12) bases his argument for life-altering change on the ethical underpinnings of the development project and the SDGs. He argues that "(c)urrent global development thinking follows the same economic growth-based version with its dominant individualist neoliberalism, overconsumption, and depoliticization of development" (p. 202), and that "(t)o transform the world to a safe, free, and environmentally sustainable one, there is a need for ethical revision of global development" (p. 202) although he does reaffirm the importance of achieving the SDGs. Yet numerous development projects have failed to deliver what they promised, and so more drastic rethinking is clearly needed if a future for all people is to be based on ethical principles that look to more than economic criteria and the assumption that growth is the answer to human ills.

Ajibade and Egge (Chapter 7) focus on three major Asian cities to emphasize the importance of tackling both issues of inequality (Goal 10) and the vulnerabilities of the poor in rapidly urbanizing societies (Goal 11). Adapting these cities to be more resilient in the face of rising sea levels and increasingly severe storm events requires tackling infrastructure, but it will be all the more effective if the social circumstances of the poor, especially those living in informal settlements, are part of adaptation planning. The record so far in Jakarta, Bangkok, and Manila suggests that much more needs to be done to make cities safe and resilient in the face of climate-related hazards and to make urban life sustainable in the long term.

Freston (Chapter 10) examines the role of religion in development, which is particularly relevant to the SDG Goals 16 on peace and 17 on partnerships. He discusses "engaging religion only where it seemed instrumental to achieving the ends pre-determined by non-religious actors" (p. 154). He argues that in addition to religious organizations' role as service provider and an influential partner for development, is also important in contributing ideas, values, and ultimately a sense of identity. He calls attention to the fundamental critiques of the ethical underpinnings of the SDGs by religious leaders, in particular the views of the Catholic Church as expressed in Laudato Si'.

Rose Taylor and Mahon (Chapter 4) focus on the comprehensive nature of the work required to incorporate gender considerations, not only in the gender goal itself (SDG 5) but mainstreamed throughout all the other goals. They argue that many contested areas require attention, ranging from sexual and reproductive rights, gender-based violence and rights of persons who identify as LGBTQ or as living with disability, as well as the more conventional gender dimensions of issues of work, pay, and property rights. A key focus in their analysis is the "spaces for contestation" in the SDG process. Rather than rejecting the (often-limited) indicators of progress thus far selected, they advocate working to broaden the indicators commensurate with the transformation required by gender mainstreaming.

The other seven chapters acknowledge the transformative nature of the agenda but focus more on how existing governance structures can be tweaked to support the SDGs. Crush's (Chapter 6) analysis of the history of the treatment of migration (a "missing" SDG) in the international agenda explains why the issue did not feature as an SDG and quoting Peter Sutherland, the UN Special Representative of the UN Secretary General on International Migration until 2017, argues that "(a)lthough Sutherland labelled the presence of migration-related issues in a handful of SDG targets a 'triumph', it was a hollow victory" (Chapter 6, p. 106). Crush does however hold out hope that migration may become more integrated into the development project with the elevation of the International Organization for Migration to the status of a UN agency in 2016, despite the often different (indeed opposing) interests of migrant-sending and migrant-receiving countries.

Whiteside's main focus (Chapter 3) is on how health became de-emphasized in the SDGs compared to its pre-eminent position in the MDGs, and the reasons for this change. He argues that the state of the HIV/AIDS epidemic played a key role. Although most of the health care system is financed domestically in most countries, he notes some global issues with implications for health governance. Three such issues are infectious disease pandemics which do not respect national borders, health of international migrants and refugees, and the impacts of climate change on health. He discusses some of the governance issues, including the fact that the key technical agency (the World Health Organization) is starved of funds, and the lack of any enforceable international framework for the right to health.

Both Horton (Chapter 13) and Weber (Chapter 14) take up the issue of funding for the SDG agenda (particularly relevant for SDG 17 on partnerships).

They discuss this from the perspective of how the necessary funds can be raised, rather than looking at governance challenges in the international financial architecture – a big topic all in itself. Horton examines how innovative sources of private finance can be mobilized for the SDGs to help to bridge the gap from "Billions to Trillions", discussing the roles of private philanthropy, blended finance, and impact investment. These have the potential of bringing in additional billions (although probably not trillions) to finance the SDGs. It seems likely that it will be hard to raise the optimistically envisaged global public funds for climate (a topic which neither Horton nor Weber discuss). The advanced countries pledged in 2015 to raise $100bn per year annually by 2020, with a major share of this to be channelled through the Green Climate Fund (Green Climate Fund 2018). However, up to May 2018, only $10.3bn had been pledged, and that includes $2bn from the US which seems unlikely to be delivered by the current administration.

Weber focuses on the only other logical place where the required volume of funds could be obtained, namely the financial industry. There are at least three different ways that banks can support the SDGs. Impact investment, where the investor is interested in a "triple bottom line" of returns (financial, social, and environmental) has the tightest link to social and environmental goals but is currently a niche product. Socially responsible investment is investment which seeks to "do no harm" and avoids financing social and environmental "bads" such as firms which flout labour laws, or the production and consumption of fossil fuels. Socially responsible investment is currently bigger than impact investment, but adopts a more passive stance vis-à-vis the SDGs, i.e. avoiding investments which harm the SDGs without specifically rewarding those which benefit them. Finally, sustainable banking takes a more proactive stance in promoting investments which help achieve the SDGs. Weber argues that although sustainable banking is not as yet large, it is growing, and he discusses the internationally developed principles and guidelines under which this can occur.

Schweizer (Chapter 9) looks at the big picture regarding potential trade-offs and synergies among the "big" goals of development, which are often categorized as economic, social, and environmental. She examines the implications of a set of modelled Shared Socio-economic Pathways (SSPs) for achievement of the SDGs, along with the associated implications for policies and institutions (i.e. international governance). Of the five SSPs modelled, she shows that the "Sustainability" pathway is the best in terms of achieving all the SDGs, whereas continuing "Historical Trends" will lead to partial achievement, while the least success occurs in the worst-case scenario ("Regional Rivalry", where national security concerns override trends towards globalization and development). There are also two second-best worlds. In "Inequality", wealthier countries and communities use technology to achieve most of the SDGs including clean energy but leave behind the poorest communities within and across countries. In "Fossil-fuelled Development", the future is mortgaged to achieve the SDGs by 2030 by increasing reliance on fossil fuels, but where after 2030 a drastic increase in carbon prices is required to halt and reverse the unsustainable rise in global temperature which ensues.

Loosely in line with such a pessimistic scenario, Quilley and Kish (Chapter 11) reject the SDGs as unsustainable, in that they are based on social and economic processes of excessive complexity, inconsistent with ecological constraints. Instead they carefully substantiate why a set of Ecologic-Economic Goals should be adopted. They envisage a very different society, one where not only the economy and society diverge considerably from the "modernist" one, but where even personality and psychology would be very different. Table 11.1 summarizes their proposed goals. Some of the differentiation factors would be the substantially more limited role of the state, disappearance of a central currency, re-emergence of the extended family, a substantial change in the role of education, and the innovation of shared participatory rituals to build social norms (since they assume that it would be very difficult to return to being a religious society). The scale of such a transformation is obviously very much greater than what is envisaged in *Transforming Our World: the 2030 Agenda for Sustainable Development* (and what the international community is currently willing to support).

The final contribution to the volume, Harrington's chapter, suggests that while there are challenges to global governance in the implementation of the goals, the process of trying to implement them inevitably will have numerous effects on global governance mechanisms. Working through numerous examples of the goals and their specific targets, she suggests that the SDGs will not only leave some practical legacies in the real world, but institutional legacies within states and international organizations. The major UN institutions as well as the World Trade Organization, the World Bank, and regional development banks are all involved in the SDGs agenda and their roles and identities are being changed in the process. While the goals have a specific timeline of the year 2030, the mechanisms designed for their implementation will undoubtedly inspire further efforts beyond that date even if, as seems likely, not all of the SDGs targets are successfully reached. In the terms of this volume's title and subtitle, what she argues is that in many ways while the SDGs are a challenge to existing global governance arrangements, they are also an opportunity to extend governance mechanisms in the future on the basis of lessons learned in the SDGs process.

The diversity of the SDGs is mirrored by the numerous perspectives in the chapters in this volume. There are no easy answers as to how to improve global governance to accomplish the SDGs. Much needs to be done in many different arenas of social and political life, and a wide variety of modes of expertise is clearly needed to facilitate the transformative changes the goals require. Read together, the chapters in this volume make the case for interdisciplinary dialogues to try to co-ordinate useful initiatives across the goals. Governance at the global scale requires new forms of research and thinking that move beyond twentieth century modes of knowledge generation where economic development could be considered apart from such things as gender equity, disease eradication, urbanization, much less climate change. The world has already been transformed in many ways. The challenge for global governance now is to guide the direction of the next stages of transformation towards a safe and sustainable future for all humanity, and for a wide range of other species too, ones whose immediate utility to the economy may not be obvious but whose ecological functions are essential for long-term sustainability for the planet.

Notes

1 The global social protection floor aimed at getting all countries to guarantee access to essential health care and basic income security for children, for older people and for those able to earn income in case of sickness, unemployment, maternity, and disability via a combination of horizontal and vertical (i.e. social security) programs. For detail on the development of the floor, a process initiated by the International Labour Organization (ILO) and its allies, see Deacon (2013).
2 These included productive employment, violence against women, social protection, inequalities, social exclusion, biodiversity, persistent malnutrition, reproductive health, peace and security, and human rights (Kamau, Chasek, and O'Connor 2018, 82).
3 Then Colombian Director of Economic, Social and Environmental Affairs.
4 Officially this was to be composed of 30 representatives chosen from the 5 UN regional groups (Africa Asia-Pacific, Eastern Europe, Latin America and the Caribbean, Western Europe and the others) but there was such interest in influencing the outcome that many resorted to a sharing of seats. This included the following "troikas": Canada Israel and the US, Australia, the Netherlands and the UK, Denmark, Ireland and Norway, France, Germany and Switzerland, Italy, Spain and Turkey (Kamau, Chasek, and O'Connor 2018, 55).
5 These were: women, children and youth, indigenous peoples, non-governmental organizations (NGOs), local authorities, workers and trade unions, business and industry, farmers, and the science and technology community. Others, including people with disabilities and migrants, were also heard.
6 A number of countries of the South would have liked to have seen the insertion of the principle of "common but differentiated responsibilities" into the SDGs but the co-chairs mentioned it in the chapeau instead (Kamau, Chasek, and O'Connor 2018, 144). Nevertheless, the principle of implementation according to national circumstances was adopted.
7 Those opposed included the UK, Australia, Canada and Switzerland (Fukuda-Parr 2019).
8 This was originally advanced by the UK Save the Children INGO (Fukuda-Parr 2019).
9 For more on the history of this issue see Rose Taylor and Mahon, Chapter 4.
10 As the SDGs went well beyond the ambitions of the MDGs, the task of identifying indicators is more complex. While some are seen to be readily available (tier 1), tier 2 are those where the concept has been agreed but measurement method needs to be elaborated, and tier 3 are those where neither the concept nor the measurement have been agreed. The task of developing indicators has been given to the Inter-Agency Expert Group (IAEG). As Fukuda-Parr (2018, 11) notes, the latter's meetings:

> subsequently have become more closed, with much of the debate now taking place in meetings limited to IAEG members (national statisticians) and the secretariat (UN statistics division). The process is not inconsequential – it is in the hands of the generalist statisticians rather than those specialized on themes.

11 See inter alia Davis, Kingsbury, and Merry (2012); Porter (2012); Broome and Quirk (2015), Kelly and Simmons (2015), and Fukuda-Parr (2014). While recognising the limitations of this governance technique, Rose Taylor (2018) argues that the answer is to critically engage.

References

Alvaredo, F., L. Chancel, T. Piketty, E. Saez, and G. Zucman. 2017. *World Inequality Report 2018*. World Inequality Lab. Accessed 2 November 2018. https://wir2018.wid.world.

Bidegain Ponte, N., and C. Rodríguez Enríquez. 2016. "Agenda 2030: A Bold Enough Framework towards Sustainable, Gender-Just Development?". *Gender and Development* 24 (1): 83–98.

Biermann, F., N. Kanie, and R. E. Kim. 2017. "Global Governance by Goal-Setting: The Novel Approach of the UN Sustainable Development Goals". *Current Opinion in Environmental Sustainability* 26–27: 26–31.

Broome, A., and J. Quirk. 2015. "Governing the World at a Distance: The Practice of Global Benchmarking". *Review of International Studies* 41 (5): 819–841.

Collier, P. 2007. *The Bottom Billion*. Oxford: Oxford University Press.

Davis, K. E., B. Kingsbury, and S. E. Merry. 2012. "Indicators as a Technique of Global Governance". *Law and Society Review* 46 (1): 71–104.

Deacon, B. 2013. *Global Social Policy in the Making: The Foundations of the Social Protection Floor*. Bristol: Policy Press.

Deacon, B. 2014. "Global and Regional Social Governance". In *Understanding Global Social Policy*, edited by N. Yeates, 53–76. 2nd edn. Bristol: Policy Press.

Esquivel, V. 2016. "Power and the Sustainable Development Goals: A Feminist Analysis". *Gender and Development* 24 (1): 9–23.

Falkner, R. 2016. "The Paris Agreement and the New Logic of International Climate Politics". *International Affairs* 92 (5): 1107–1125.

Fukuda-Parr, S. 2014. "Global Goals as a Policy Tool: Intended and Unintended Consequences". *Journal of Human Development and Capabilities* 15 (2–3): 118–131.

Fukuda-Parr, S. 2018. "Sustainable Development Goals (SDGs) and Global Goals". In *Oxford Handbook on the United Nations, Oxford Handbooks Online*, edited by S. Daws and T. G. Weiss, 1–17. 2nd edn. Oxford: Oxford University Press. DOI: 10.1093/oxfordhb/9780198803164.013.42.

Fukuda-Parr, S. 2019. "Keeping Out Extreme Inequality form the SDG Agenda – The Politics of Indicators". *Global Policy* 10 (Supplement 1): 61–69.

Fukuda-Parr, S., and D. Hulme. 2011. "International Norm Dynamics and 'The End of Poverty': Understanding the Millennium Development Goals". *Global Governance* 17 (1): 17–36.

Fukuda-Parr, S., and T. S. Hegstad. 2018. "'Leaving No One Behind' as a Site of Contestation and Reinterpretation". *Department of Economic and Social Affairs Background Paper 47* ST/ESA/2018/CDP/47.

Fukuda-Parr, S., and D. McNeill. forthcoming. "Knowledge and Politics in Setting and Measuring the SDGs: Introduction'". *Global Policy* 10 (Supplement 1): 5–15.

Fukuda-Parr, S., A. E. Yamin, and J. Greenstein. 2014. "The Power of Numbers: A Critical Review of MDG Targets on Human Development and Human Rights". *Journal of Human Development and Capabilities* 15 (2–3): 105–117.

Green Climate Fund. 2018. "How We Work: Resource Mobilization". *Green Climate Fund*. Accessed 1 November 2018. www.greenclimate.fund/how-we-work/resource-mobilization.

High-Level Panel of Eminent Persons on the Post-2015 Development Agenda. 2013. *A New Global Partnership: Eradicate Poverty and Transform Economies through Sustainable Development*. New York: United Nations.

ICIJ (International Consortium of Investigative Journalists). 2017a. *Paradise Papers: Secrets of the Global Elite*. Accessed 19 November 2018. www.icij.org/investigations/paradise-papers/.

ICIJ (International Consortium of Investigative Journalists). 2017b. *The Panama Papers: Exposing the Rogue Offshore Finance Industry*. Accessed 19 November 2018. www.icij.org/investigations/panama-papers/.

Kahn, J. 2018. "Google's 'Dutch Sandwich' Shielded 16 Billion Euros from Tax". *Bloomberg*, 2 January. Accessed 19 November 2018. www.bloomberg.com/news/articles/2018-01-02/google-s-dutch-sandwich-shielded-16-billion-euros-from-tax.

Kamau, M., P. Chasek, and D. O'Connor. 2018. *Transforming Multilateral Diplomacy: The Inside Story of the Sustainable Development Goals*. London: Routledge.

Kelly, J. G., and B. A. Simmons. 2015. "Politics by Number: Indicators as Social Pressure in International Relations". *American Journal of Political Science* 59 (1): 55–70.

Lakner, C., and B. Milanovic. 2013. "Global Income Distribution: From the Fall of the Berlin Wall to the Great Recession". *Policy Research Working Paper* No. 6719. Washington DC: World Bank.

Porter, T. 2012. "Making Serious Measure: Numerical Indices, Peer Review and Transnational Actor-Networks". *Journal of International Relations and Development* 15 (4): 532–557.

OECD (Organisation for Economic Co-Operation and Development). 2008. *Growing Unequal: Income Distribution and Poverty in OECD Countries*. Paris: OECD.

OECD (Organisation for Economic Co-Operation and Development). 2011. *Divided We Stand: Why Inequality Keeps Rising*. Paris: OECD.

OECD (Organisation for Economic Co-Operation and Development). 2018. "Base Erosion and Profit Shifting". *OECD*. Accessed 19 November 2018. www.oecd.org/tax/beps/.

Piketty, T. 2014. *Capital in the Twenty-first Century*. Cambridge, MA: Harvard University Press.

Razavi, S. 2016. "The 2030 Agenda: Challenges of Implementation to Attain Gender Equality and Women's Rights". *Gender and Development* 24 (1): 25–41.

Rose Taylor, S. 2018. "On Measurement and Meaning: How Indicators Shape, Make, or Break Global Policy Goals and Outcomes". *The Role of Indicators in Promoting Gender Equality Through the Millennium and Sustainable Development Goals*, PhD diss., Wilfrid Laurier University.

Sachs, J., G. Schmidt-Traub, C. Kroll, G. Lafortune, and G. Fuller. 2018. *SDG Index and Dashboards Report 2018*. New York: Bertelsmann Stiftung and Sustainable Development Solutions Network (SDSN).

Selby, J. 2018. "The Trump Presidency, Climate Change, and the Prospect of a Disorderly Energy Transition". *Review of International Studies* 1–20.

UNDESA (United Nations Department of Economic and Social Affairs). 2012. *Realizing the Future We Want for All*. Report to the Secretary General of the UN Task Team on Post-2015 Development Agenda. New York: UN.

UNGA (United Nations General Assembly). 2015. *Transforming Our World: the 2030 Agenda for Sustainable Development*. A/RES/70/1. Accessed 17 November 2018. https://undocs.org/A/RES/70/1.

United Nations. 2015. *Millennium Development Report 2015*. Geneva: UN Secretariat. Accessed 5 December 2018. www.un.org/millenniumgoals/2015_MDG_Report/pdf/MDG%202015%20rev%20(July%201).pdf.

Unterhalter, E. 2019. "The Many Meanings of Quality Education: Politics of Targets and Indicators in SDG4". *Global Policy* 10 (Supplement 1): 39–51.

World Bank. 2013. *A Common Vision for the World Bank Group*. DC2013-0002, 3 April. Accessed 2 November 2018. http://siteresources.worldbank.org/DEVCOMMINT/Documentation/23394965/DC2013-0002(E)CommonVision.pdf.

Zürn, M. 2018. *A Theory of Global Governance: Authority, Legitimacy and Contestation*. Oxford: Oxford University Press.

2 Food system lessons from the SDGs

Alison Blay-Palmer and Laine Young

(Un)Sustainability through the lens of food

Food is both a significant contributor to global challenges as a well as a potential pivot point for transition and transformation as we tackle climate change, global health challenges such as obesity/overweight and diabetes, growing mountains of waste, declining water quality/quantity, and renewable energy availability (Kloppenburg *et al.* 2000; Sonnino and Marsden 2005; Seyfang 2006; McMichael 2011). Addressing food insecurity, diet-related illness, food waste, and the energy used in the food system could have a huge positive impact on global development. Including gender into this analysis allows us to see how food system issues are impacted by intersections of inequality and provides insights into how the Sustainable Development Goals (SDGs) can help improve livelihoods.

Food security is a significant challenge we continue to grapple with as a world. Defined as "… when all people, at all times, have physical, social and economic access to sufficient, safe and nutritious food to meet their dietary needs and food preferences for an active and healthy life" (FAO 2006, 1), the goal of food security for all continues to elude us. Despite having more than enough food to feed every person on the planet, hundreds of millions of people experience food insecurity and the numbers are on the rise again. Globally, food insecurity increased by 38 million people from 2015 to 815 million people in 2016 (FAO *et al.* 2017). The United Nations Food and Agriculture Organization (FAO) suggests that the increasing trend is linked to the singular and compounding effects of conflicts and climate change with notable changes in sub-Saharan Africa, South-Eastern Asia, and Western Asia. Children who are malnourished will not reach their full potential as adults given this deprivation. While on the decline, there are still 155 million children suffering from stunting and 52 million who suffer from wasting (FAO *et al.* 2017). In addition, the nutrition transition is taking a toll as children globally experience increasing levels of obesity. Between 2005 and 2016, the percentage increased from 5.3 to 6.0 per cent, an increase of 13.2 per cent. This number is especially concerning as increases are occurring in Africa (from 5.0 to 5.2, a 4 per cent increase), Asia (4.4 to 5.5, a 25 per cent increase) and Latin America and the Caribbean, from 6.8 to 7.0, a 3 per cent increase).

In the general population, the World Health Organization (WHO 2017a) reported that obesity has nearly tripled since 1975 so that in 2016 there were more 1.9 million overweight adults (39 per cent) including 650 million (13 per cent) who were obese. This translates into increased diet-related mortality as currently, "Most of the world's population live in countries where overweight and obesity kills more people than underweight" (WHO 2017a). Food-related chronic disease and death is also on the rise. Between 1980 and 2014 global diabetes rates nearly quadrupled to 422 million people with an increase to 8.5 per cent from 4.7 per cent with the most rapid increases in middle- and low-income countries. In 2015, it is estimated that 1.6 million deaths were directly caused by diabetes (Mathers and Loncar 2006; WHO 2017b). In addition to these health-related impacts, negative environmental consequences are also linked to the industrial food system.

The global food system, from fertilizer manufacture to food storage and packaging, is responsible for up to one-third of all human-caused greenhouse gas (GHG) emissions. According to the latest figures from the Consultative Group on International Agricultural Research (CGIAR), a partnership of 15 research centres around the world, "… reducing agriculture's carbon footprint is central to limiting climate change. And to help to ensure food security, farmers across the globe will probably have to switch to cultivating more climate-hardy crops and farming practices" (Vermeulen *et al.* 2012 as quoted in Gilbert 2012). Breaking this down, livestock production contributes about two-thirds to this total (FAO 2018). Regionally, up to 86 per cent of the total GHG emissions are from food production, although this varies based on production practices. In developed countries post-production contributes the biggest share while in other countries, e.g. Sri Lanka, food handling and distribution are the greatest sources of waste (Dubbeling, Carey, and Hochberg 2016).

Looking at the equation from the perspective of food production and availability, impacts from climate change on the food system are predicted to result in higher temperatures and more flooding. A report by Thornton estimated the impacts of climate change on 22 crops so that "By 2050, climate change could cause irrigated wheat yields in developing countries to drop by 13%, and irrigated rice could fall by 15%. In Africa, maize yields could drop by 10–20% over the same time frame" (as quoted in Gilbert 2012). The concluding recommendations from the research are:

> For some crops, improvements to heat resistance through conventional and transgenic breeding, for example, will help farmers to adapt. But for others, more radical changes are needed. Thornton says that potato-growing areas, including China and India, are likely to see yields drop significantly as temperatures rise, and he suggests that farmers consider growing crops, such as bananas, that do better in warmer climates.
>
> Campbell says that the CGIAR will use the paper to help set its research agenda for the next decade and "identify which crops and which regions to focus investment on".

He calls on governments meeting next month at the climate-change conference in Doha, Qatar, to agree on a way forward to tackle the challenges of mitigating and adapting to the effects of climate change on agriculture.

(As quoted in Gilbert 2012)

These recommendations are worrisome as they point in the direction of high-technology solutions such as what is called "Climate Smart Agriculture" and "Sustainable Intensification", both of which would rely on increased chemical inputs innovation such as transgenic breeding. This approach discounts the value of traditional farming knowledge and practice, and in particular the role women play in food systems. In addition to decreased yields, it is also expected that climate change effects will impact food safety with increases in the risk of food-borne illnesses and diarrheal diseases.

In addition to individual and household food insecurity and well-being and climate change impacts, food is also linked to water availability and usage. On the production side, agriculture uses up to 70 per cent of global water withdrawals (United Nations 2014; FAO 2016) while food production and its supply chain use approximately 30 per cent of global energy (UNESCO 2012). Food production can be water intensive. For example, to produce 1 kg of rice requires 3,000 to 5,000 litres of water, 1 kg of soya requires 2,000 litres, 900 litres are needed for wheat and 500 litres for potatoes (Schuyt and Brander 2004). A further consideration is the degradation and diminishing of ground water supplies; it is estimated that 20 per cent of groundwater supplies are over-exploited. Wetlands are also in jeopardy with an associated diminished capacity for ecosystem water filtration (UNESCO 2014).

Given the intensive resources needed to produce food, one would assume that food waste would be as minimal as possible. Unfortunately, this is not the case. It is estimated that one-third of food is wasted globally amounting to about 1.3 billion pounds every year across the food supply chain (FAO 2011). This is especially concerning given the resources used to produce, process, and transport including soil, land, energy, water, and associated GHG emissions. The SDGs are an attempt to address these equity and resource management challenges in a more integrated way.

The Sustainable Development Goals

The United Nations describe the rationale for and the purpose of the SDGs as "to build on the success of the Millennium Development Goals (MDGs) and aim to go further to end all forms of poverty" (UN-SDG 2017). While the success of the MDGs is contested (e.g. Fehling, Nelson, and Venkatapuram 2013), in the spirit of building forward from the MDGs towards the SDGs, Griggs and his colleagues (2013) encouraged us to adjust the concept of sustainable development to one where the environment is the all-encompassing container with society and the economy nested within. Given this reframing we are aiming for "development that meets the needs of the present while safeguarding Earth's life-support system, on which the welfare of current and future generations depends" (Griggs *et al.* 2013, 306). This shifts the focus from the well-known three overlapping rings with one for each of the economy, the environment, and society to a more deliberate acknowledgement of the foundational place of the environment to planetary and human well-being so that the economy and society are nested within the environment.

Since the SDGs adoption on 1 January 2016, countries are working to interpret, measure, and act on them. There have been three rounds of voluntary reviews – in 2016, 2017, and 2018 – including 22 countries in 2016, 43 in 2017, and 46 in 2018. To date, these Voluntary National Reviews (VNRs) are presented to the High Level Political Forum on Sustainable Development annually, and they:

> aim to facilitate the sharing of experiences, including successes, challenges and lessons learned, with a view to accelerating the implementation of the 2030 Agenda. The VNRs also seek to strengthen policies and institutions of governments and to mobilize multi-stakeholder support and partnerships for the implementation of the Sustainable Development Goals.
> (Sustainable Development Knowledge Platform 2018)

The basis for the VNRs is the 17 SDGs and their associated 167 indicators.

While applauded as a multi-lateral achievement (Deacon 2016), given the increasing focus on using the SDGs, it is important to review ways they have been critiqued. First, they are seen as technocratic, as relying heavily on data and Western science, and thus as privileging Western science over place-based knowledge systems, which contributes to the further marginalization of the most disadvantaged, including women, the elderly, and youth especially in the Global South (CFS 2015). Further, in some developed countries, the application of the SDGs to countries in the Global South and not the Global North is similar to the MDGs. For example, in the UK a recent Environmental Audit Committee report pointed out that while the SDGs provide an opportunity to develop sustainability-focused policy coherence within the UK, the approach within the government is either that the UK's role is to help other countries with their adoption, thereby missing possibilities in the UK, or for their uptake to be one-off, resulting in a disjointed approach. The report proposes that to remedy this situation, the UK government should demonstrate leadership with respect to the SDGs including an independent advisory committee, a Cabinet-level minister, as well as regular reporting beginning in 2018 with the VNR (Benson-Whalén 2017).

Scholars at Yale question the extent to which the SDGs slow, rather than accelerate, the realization of the human right to food. There are political realities that must be acknowledged in trying to achieve the SDGs, including the dominance of short-termism (Morley and Marsden 2014). For instance, the target for reducing inequality is 2029. Nevertheless, Pogge and Sengupta (2016) suggest there may be lags in expediently addressing inequity, leaving the powerful time to undertake further capture of wealth and resources. They also point to a lack of understanding as to who does what, so if the SDGs or elements of the goals fail, there will be no mechanism to understand why or how to recover (Pogge and Sengupta 2016).

The SDGs and food

With these cautions in mind, many countries have taken up the challenge of addressing the SDGs and look to them to enable sustainable development. Within this context, food has the potential to play a significant role. The second SDG, to "end hunger, achieve food security and improved nutrition, and promote sustainable agriculture" (UNGA 2015, 14) includes sub-goals with target dates from 2025 to 2030.

It is important to pay attention to the intended targets as the SDGs were developed in 2014 and the final indicators were inserted later. When the two are compared, it is clear that there has been a narrowing of scope and thus also expectations. This is inevitable to the extent that the SDGs, like the MDGs before them, rely on that which is more readily measured. For example, the first indicator, prevalence of undernourishment as an annualized average, does not capture seasonal variation inherent to food systems and so misses periods in the year when people do not have enough food. In the case of Target 2.4, there is also a disconnect between the intention of developing more resilient and sustainable food systems and the single indicator of "Proportion of agricultural area under productive and sustainable agriculture". While this indicator is being developed by the FAO to include information gathered at the farm scale and to include the three pillars of sustainability (FAO 2017), this type of data gathering can be expensive and as a result may preclude the participation of the least developed countries.

A coherent approach to sustainability through food? Points of intersection between the goals

Certainly, while there is interest in developing more inclusive holistic approaches to sustainable development, the SDGs indicators in some ways limit this possibility. That said, and acknowledging these limitations, food offers a cross-cutting and coherent way to tackle the challenges identified through the SDGs. One way to create coherence across regional, national, and global policies and conventions, like the SDGs, is a City Region Food Systems (CRFS) approach (Blay-Palmer *et al.* 2018). This approach assesses the sustainability of a city region with regards to flows of resources, such as knowledge, people, food, and waste between rural, peri-urban, and urban areas, and the necessary policies and procedures to facilitate a sustainable food system in the region. It is also important to consider the linkages between SDG 2 and other goals.

Table 2.1 Intent vs. reality of operationalizing the goals

Intended target, UN-2014	SDG indicator (FAO 2018)
2.1 Ending hunger for all people including infants year-round;	Prevalence of undernourishment
2.2 Ending all kinds of malnutrition with particular attention to the elderly, adolescent girls, pregnant and lactating women and with stunting and wasting ended by 2025;	Prevalence of moderate or severe food insecurity in the population, based on the Food Insecurity Experience Scale (FIES)
2.3 Double agricultural productivity and smallholder producer incomes with a focus on women, indigenous peoples, family farmers, pastoralists and fishers through secure land access and equal access to productive resources, and opportunities for value-added and off-farm employment;	Volume of production per labour unit by classes of farming/pastoral/ forestry enterprise size Average income of small-scale food producers, by sex and indigenous status
2.4 Ensure sustainable production and agricultural practices that support resilience, ecosystems, climate change adaptation including extreme weather, and that improve land and soil;	Proportion of agricultural area under productive and sustainable agriculture
2.5 By 2020 develop regional, national and international well-managed, diverse seed and animal gene banks, and equitably share benefits from use of genetic resources and traditional knowledge;	Number of plant and animal genetic resources for food and agriculture secured in medium- or long-term conservation facilities Proportion of local breeds, classified as being at risk, not-at-risk or unknown level of risk of extinction
2.6 Provide more investment in rural infrastructure including gene banks and technology development, research and extension in developing, especially least developed, countries; address and correct distorting international trade barriers in agricultural markets including the elimination of export subsidies; and, address dysfunctional food commodity and derivative markets including access to market information and food reserves to eliminate volatile food prices	The agriculture orientation index for government expenditures Indicator of (food) price anomalies

Source: Alison Blay-Palmer.

Various research has demonstrated that food can act as both a unifying cross-cutting theme connecting the SDGs and indicators. Building on work by the International Social Sciences Council (ISSC), a mapping initiative was undertaken to better understand the linkages between the various dimensions that are the targets for the SDG indicators. Based on this assessment, hunger, climate change, energy, and water emerge as four foci central to addressing the challenges identified by the SGDs. Food adds to (or can be seen as addressing) these challenges. Food can also act as a transformational nexus point across the SDGs, where enhanced connections address the goals to end poverty and hunger, improve employment opportunities, build scale-appropriate infrastructure, foster social inclusion and sustainability, and prioritize gender equality and empowerment (Figure 2.1 from Blay-Palmer *et al.* 2018). Food provides an entry point for multiple scales including cities and their surrounding regions to engage in coherent regional food policy and programme initiatives that focus on governance, sustainable diets, and nutrition, social and economic equity and food production, distribution, waste reduction, and recovery (Dubbeling, Carey, and Hochberg 2016; Forster *et al.* 2015; Dubbeling *et al.* 2017).

The Voluntary National Review process points to the benefits that could accrue in using a food lens. For instance, according to the review of the 2017 VNR process:

> Prominent in the coverage of the SDG 2 on food security were the inter-linkages with other SDGs, particularly poverty eradication, as well as job creation and women's empowerment. Several countries reported on putting

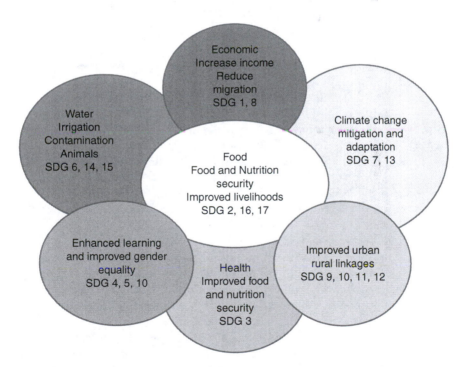

Figure 2.1 Potential for food as an integrative approach to addressing SDGs.

Source: reproduced with permission from Alison Blay-Palmer, Guido Santini, Marielle Dubbeling, Henk Renting, Makiko Taguchi, and Thierry Giordano, Sustainability; published by Molecular Diversity Preservation International (MDPI), 2018.

legal and policy frameworks in place to ensure the right to food as well as initiatives aimed at ending hunger and food insecurity. Examples ranged from national plans and policies on food security to more specific initiatives such as food security programmes for society's most vulnerable groups. Many countries reported on challenges as well as initiatives to ensure sustainable food production systems, since ending hunger and malnutrition relies heavily on sustainable food production systems and resilient agricultural practices. Countries are also concerned about the impact of climate change on agriculture, as well as consequences of natural disasters for agriculture and food security.

(UNDESA 2017, vii)

Food system relevant observations from the 2017 review emerged from across the goals and include:

- SDG 1 (End Poverty) and the need to specifically address marginalization and exclusion, with particular attention required to address child poverty, resources of single parent families, and youth employment. Addressing this would also help to ensure access to food and could decrease food insecurity.
- For SDG 3 (Ensure healthy lives and promote well-being for all at all ages), shared initiatives included adopting the constitutional right to health where this has been implemented in several countries including Ethiopia, Kenya, Nepal, Portugal, and Zimbabwe (United Nations 2017, 9). As adequate food is a prerequisite for health, the increase in non-communicable disease was a notable concern with specific references to diabetes and cardio-vascular disease, and attention to healthy eating and lifestyle initiatives (see Whiteside, Chapter 3).
- Other food-connected dimensions include SDG 4 and the opportunities for education especially for girls, and gender empowerment as part of SDG 5 (Gender Equality) to address land access and land tenure rights as well as fair access to markets. Educational opportunities provide opportunities for income generation and also empower women and girls to be more engaged in their communities, which can lead to better food security (see Rose Taylor and Mahon, Chapter 4).
- SDG 6 (Clean Water and Sanitation) identified challenges included access to clean drinking water particularly for rural communities, mounting pressures on water resources for competing needs including food production, and the links between water and biodiversity.
- SDG 7 (Affordable and Clean Energy) connects with sustainable food systems through calls for attention to better rural connectivity, as well as energy demand reductions that could be addressed, for example, by decreasing food waste from a more closed loop, efficient food system.
- SDG 8 (Decent Work and Economic Growth) discusses noted employment challenges for youth and the elderly, as well initiatives to support

small- and medium-size businesses, potentially food system-related, and community-driven economic activity. A focus in this area was both on investment through financial means as well as human capital.

- Many dimensions of SDG 9 (Industry, Innovation, and Infrastructure) are relevant in the food context, including improved transportation networks, for example, roads, access to clean water, irrigation, and power supplies.
- SDG 10 (Reduced Inequalities) can be addressed through the integrative dimensions of food through cultural inclusion as well as opportunities for youth and women empowerment.
- Results reported specific to SDG 11 that have a food orientation are the need for infrastructure and improved mobility to support urban settlements and the growing air, water, and soil quality pressures from urbanization.
- For SDG 12 (Responsible Production and Consumption) developing green and circular economic activity with a focus on food waste reduction and green procurement as well as opportunities for consumer education are particularly noteworthy.
- SDG 13 (Climate Action) addresses reported changes including flooding, rain variability, and extreme weather with solutions including ecosystem-based adaptation and carbon-sinks, carbon-pricing, cross-border cooperation on disaster reduction, flood management, and weather forecasting with associated benefits, including economic benefits from low-carbon or no-carbon societies, such as fewer emissions, increased competitiveness, and growth and employment (See Dalby, Chapter 8).
- SDG 14 (Life Below Water) is clearly linked to food system considerations as a food source and part of biodiversity. Collective insights of import include addressing pollution, overfishing, and illegal fishing with the recognition for more financial and institutional resources to deal with these problems.
- SDG 15 (Life on Land) resulted in similar concerns as the previous goal including illegal activity and preserving biodiversity integrity with solutions requiring more robust institutions, including environmental laws.
- Applying a food lens to the findings for SDG 16 (Peace, Justice and Strong Institutions) and SDG 17 (Partnerships for the Goals), necessary conditions include the rule of law and the respect for human rights, including the call to respect official development assistance as a stabilizing mechanism. Common threads in many of the SDGs reports were dealing with inequity, establishing institutional supports, stable resources, including funding to countries in the Global South through Official Development Assistance (ODA), and dealing with climate change. All of this can be addressed in some part through more sustainable food systems, as will be discussed in the following section on gender through urban agriculture and agroecology.

(UNDESA 2017, vii–xi)

Adopting an increasingly coherent approach to the SDGs would offer the opportunity to enhance sustainability. However, doing this requires a level of policy coherence so that contradictions and negative feedback loops are avoided. For example, in addressing food waste by developing market flows that are less wasteful, it is important to ensure that employment opportunities and food access, especially for youth and women, are not compromised.

Links between SDG 2 and SDG 5

Like Goal 2, Goal 5 (achieve gender equality and empower all women and girls) can be integrated into many of the other SDGs (see Rose Taylor and Mahon, Chapter 4). The links between Goals 5 and 2 are critical to consider when implementing a more integrated approach to meeting SDGs targets. Gender adds an additional element to Goal 2 by examining inequality and unequal power relations within food security, food access, nutrition, agricultural productivity, preservation of genetic diversity, trade restrictions, as well as market and land access. Of course, when applying gendered frameworks to development, an intersectional approach is critical. Intersectionality, as proposed by Crenshaw in 1989, rejects the essentialism of women's experience and recognizes that different social categories (race, gender, class, etc.) are separate experiences that intersect and affect an individual's participation in society.

When it comes to ending hunger and malnutrition, gender is a critical element to consider because of specialized nutritional needs (pregnancy, lactation, etc.), as well as the gender roles present in much of the world that put women as the primary responsible for food provisioning (Castellano 2015). Because it is regularly women's job to ensure the food security of the home, the pressure is significant. Often women will go without food or consume much less to ensure the rest of their family has as much as is available to them. Another target within Goal 2 is to double agricultural productivity and incomes of small-scale producers with a special focus on vulnerable populations like women. It is well known that women are impacted by inequality in access to land and other resources, inputs, financial services, knowledge, markets, and opportunities for value addition and other employment. The United Nations Development Programme has shown that inclusion of gender is integral in assessing these targets. Moreover, in considering gender and other intersectional elements that will impact implementation of SDG 2, it is critical to ensure that those who are most vulnerable to the effects of food insecurity are represented in the implementation plan.

Some examples of the importance of inclusion of intersectional experiences in meeting these indicators are as follows. As women are especially vulnerable to disasters, the target that promotes sustainable food production systems and resilient practices that can improve capacity of adaption to climate change and disasters also requires an analysis of gender with other factors like class and ethnicity. The target related to genetic diversity of seeds is impacted by these intersections as a large amount of small-holder women farmers are seed saving and using traditional knowledge to feed themselves and their community

(Wichterich 2015). In Taylor (2018) the critical discussion of Climate Smart Agriculture (CSA) indicates the potential increase in vulnerability for women.

Proponents of CSA, like the World Bank and the FAO, believe it is necessary to implement this heightened production with increased resilience to climate change for the continued supply of food for the world's population and for the health of our environment. Taylor (2018) argues that while CSA policies are described as holistic, they lack socio-political context and focus primarily on technical aspects of production. This leads to a lack of focus on power, access, and inequality (Taylor 2018) which is critical when moving towards solutions to food security and climate change. "Put simply, CSA fails to address how enduring inequalities of access in both production and consumption strongly shape who is rendered vulnerable to climate change and who is left food insecure" (Taylor 2018, 90). Additionally, working towards correction and prevention of trade restrictions and distortions affects women because moving from growing traditional food to food for the global market often affects their capacity to feed the family and moves control of growing to the men in the household, as they are often in charge of the market. Women lose control of their food provisioning and the household income from food growth. Finally, proper functioning of markets can be analysed through a gendered lens, as in some areas women are responsible for market sales of food for the family, and access to information and knowledge is often not passed on to women. These impacts need to be considered when assessing these targets as well as designing implementation of solutions.

Urban agriculture and agroecology

Urban agriculture (UA) is a practice being implemented by many cities to promote food security, increase income and employment, provide environmental benefits, and increase diversity in consumption (Prain and Dubbeling 2011). UA has the potential to assist with meeting the following targets within Goal 2: end hunger by ensuring access to food, end malnutrition (promoting maternal and infant health), promote agricultural productivity of small-scale food producers, promote resilience to disasters, maintain genetic diversity, promote increased investment through international co-operation and research, compete with the globalized food system, and affect the food market positively. However, UA is greatly affected by differential access to resources, socio-economic status, and social and cultural norms can all affect the experience of growing food in the city. This practice has many inequalities present that can lessen the positive impact on reducing hunger and other food-related goals. Inequalities in UA are related to differences in access to power in the home, community, and larger scales. Many suggest that the majority (65 per cent) of people participating in UA globally are women (Dubbeling, de Zeuw, and van Veenhuizen 2011). The compounding intersectional effects of one's gender, race, class, and other social locations significantly affect access to UA and therefore food security and economic independence. When working to meet the goals from SDG 2 it is crucial that an intersectional analysis of access is considered when moving towards solutions, like UA.

The addition through food sovereignty of democratic engagement and community participation is a necessary complement to the recognition of the links between people as part of the food ecosystem.

> While in existence for millennia, for the last two decades agroecology is closely linked to principles of food sovereignty. These principles put the Right to Food as the foundational consideration and the centrality of people and not states as first and foremost. They also include the need for: agrarian reform so that women, landless and indigenous people own and control the land they work; that natural resources, especially land, water, seeds and livestock breeds are cared for sustainably including a move away from agro-chemical use towards balanced and diversified ecosystems and that are able to use freely and to protect the genetic resources communities have developed through their history.
>
> (Pimbert 2017, 340)

> [F]ood production should focus on domestic consumption first and trade second; control of the commodification and speculation of food; food as a source of peace not violence; and that smallholder farmers have venues to provide input into policies at all scales.
>
> (Via Campesina 1996 as quoted in Pimbert, 2017, 340–367)

Given these foci, a compelling question in the context of SDG 2, the goal to end hunger, would be whether a shift from the dominant industrial, conventional food systems approach to one based on agroecology would help to address the SDGs. To consider this question, four interconnected areas will be addressed: food production, consumption, and food security (SGDs 2 and 12); ecosystem well-being and biodiversity (SDGs 14 and 15); climate change pressures (SDG 13); and gender equality (SDG 5).

In terms of food production as the underpinning for providing enough food, while qualified, there is a strong consensus that adopting agroecology could produce enough to feed the world. In a review of research into crop productivity in organic versus conventional agriculture, results support the ability of organic farming to produce approximately 80 per cent of comparable amounts of food when one looks across the food system, including commodity-based crops such as maize and soybeans (De Ponti *et al.* 2012; Seufert, Ramankutt, and Foley 2012). Agroecology as a more comprehensive food system would offer adequate yields and also community and health benefits. Given that food security is about both having enough quality food *and* access to that food, the amount of food produced is only one dimension of the food security equation addressed by SDG 2. According to the International Assessment of Agricultural Knowledge, Science and Technology for Development (IAASTD) report, "… a focus on small-scale sustainable agriculture, locally adapted seed and ecological farming

[can] better address the complexities of climate change, hunger, poverty and productive demands on agriculture in the developing world" (McIntyre 2009).

When we consider the benefits of addressing climate change through agro-ecology the calculation becomes even more interesting. As indicated in previous sections, current food system practices are huge climate change culprits. Practicing agriculture in a more ecological manner offers ways to limit climate change impacts across the input spectrum, and it could specifically help address the need:

> ... to limit nitrogen and phosphorus use in agriculture. Nutrient use efficiency should improve by 20% by 2020; no more than 35 million tonnes of nitrogen per year should be extracted from the atmosphere; phosphorus flow to the oceans should not exceed 10 million tonnes a year; and phosphorus runoff to lakes and rivers should halve by 2030.
>
> (Griggs *et al.* 2013, 307)

In addition, permaculture and agro-forestry provide ways to reduce and draw down carbon dioxide, reduce soil erosion and rebuild soil, provide resistance to droughts, and help mitigate flooding (Altieri and Koohafkan 2008).

Gender is an important aspect to consider within agroecology. As women's roles in the household are often undervalued and they have less decision-making capacity in the home, the right to food is complicated. If programs are targeting households, they may not be reaching the women. Often men are seen as the heads of household and the main agricultural workers leading to their title of land owner (Schwendler and Thompson 2017). Schwendler and Thompson (2017) suggest that education empowers women, which can increase their participation in agroecology. Not only this, women's education in agroecology and politics is critical to transforming unequal gender relations. The authors describe the Landless Rural Workers Movement (MST) program in Brazil where feminism was integrated into their movement to increase education and combat the patriarchal structures present in their community.

Brazil's MST provides education in agroecology in communities. It uses a pedagogical approach that is successful because of its inclusion of (1) debates and discussions about gender integrated into education curriculum; (2) the involvement of both men and women to challenge the division of labour, and (3) increasing women's access to technical knowledge to participate in agroecology (Schwendler and Thompson 2017). On the community side of the equation, agroecology also offers community-building opportunities through knowledge sharing within and between communities as part of social movement building. The international La Via Campesina is an excellent example of the important role agroecology plays in this regard. In addition, there is a need to "theorize up" and move beyond anecdotal case study work that has taken place mostly to date to understanding wider benefits (Moragues-Faus and Marsden 2017).

For food system changes to take place, it is important not to subsume agro-ecology into the industrial food system but for it to activate transformative change as a social movement (Holt-Giménez and Altieri 2013). Moving towards agroecology would require significant structural changes. Horlings and Marsden point to multiple considerations that would need to be addressed for agroecology to succeed, so that while:

> agro-ecological approaches could significantly contribute to "feeding the world", and thereby contribute to a "real green revolution"; ... this requires a more radical move towards a new type of regionally embedded agri-food eco-economy. This is one which includes re-thinking market mechanisms and organisations, an altered institutional context, and is interwoven with active farmers and consumers' participation. It also requires a re-direction of science investments to take account of translating often isolated cases of good practice into mainstream agri-food movements.
>
> (Horlings and Marsden 2011, 441)

In the context of the SDGs and the facilitative role agroecology can play in addressing multiple global challenges, it is important to recall that the SDGs are voluntary so need to offer opportunities that warrant data collection and report writing associated with the Voluntary National Review and related processes. Given the aspirational and voluntary status of the SDGs, the VNRs have offered countries the opportunity to share their process for interpreting and applying the SDGs and the associated targets, their accomplishments and lessons learned. While it is understood that the application and uptake of the indicators is nationally and sub-nationally specific, the VNRs provide a learning space where countries can share their knowledge. This can act as a way to enable the emergence of a common set of global governance tools and processes that can be adopted in place-based ways. Adopting agroecology as much as possible could be one step in that process.

Conclusion: linking the SDGs with other policy mechanisms

In addition to the SDGs, there are two other key policy mechanisms that are giving more prominence to food and the way it can act as a lever for change. Both emerged from urban initiatives but also speak to the importance of territorial considerations. The first, the Milan Urban Food Policy Pact (MUFPP) is a voluntary set of sustainability guidelines that provides a framework for change. In addition, the indicators are being modified to include gender as an important lens for cities to assess their progress across. This focus on the importance of gender data disaggregation in the MUFPP will allow for a more intersectional assessment of the sustainability of each city involved. For example, when assessing the vulnerability of a city to disasters, SDG 11 discusses reducing the amount of deaths and people affected by disasters by focusing on the poor and other vulnerable populations. The MUFPP includes an indicator on the existence of a municipal food resiliency plan in the face of events like disasters. We must also consider how gender can affect the vulnerability to disasters, and what needs to be done to ensure equality in this area. The New Urban Agenda also points to food as a platform for change in the context of climate change. Particular reference is made to increasing food security, reducing food waste (an important contributor to GHG emissions including methane), and improve resource use, including water. While all of these initiatives were adopted since 2015, they are framing discussions about how to build sustainability from the city region through to international discussions and agreements.

References

Altieri, M. A., and P. Koohafkan. 2008. *Enduring Farms: Climate Change, Smallholders and Traditional Farming Communities.* Vol. 6 of Environment and Development Series. Penang: Third World Network (TWN).

Benson-Wahlén, C. 2017. "UK Committee Critiques Approach to SDGs". *IISD*, 2 May. Accessed 26 November 2018. http://sdg.iisd.org/news/uk-committee-critiques-approach-to-sdgs/.

Blay-Palmer, A., G. Santini, M. Dubbeling, H. Renting, M. Taguchi, and T. Giordano. 2018. "Validating the City Region Food System Approach: Enacting Inclusive, Trans-formational City Region Food Systems". *Sustainability* 10 (1680): 1–23.

Castellano, R. L. S. 2015. "Alternative Food Networks and Food Provisioning as a Gendered Act". *Agriculture and Human Values* 32 (3): 461–474.

CFS. 2015. *Civil Society Statements to CFS 42.* Accessed 26 November 2018. www.csm4cfs.org/wp-content/uploads/2016/04/En-CS-contribution-2015-BR.pdf.

Crenshaw, K. 1989. "Demarginalizing the Intersection of Race and Sex: A Black Feminist Critique of Antidiscrimination Doctrine, Feminist Theory and Antiracist Politics". *University of Chicago Legal Forum* 1: 139–167.

De Ponti, T., B. Rijk, and M. K. Van Ittersum. 2012. "The Crop Yield Gap Between Organic and Conventional Agriculture". *Agricultural Systems* 108: 1–9.

Deacon, B. 2016. "SDGs, Agenda 2030 and the Prospects for Transformative Social Policy and Social Development". *Journal of International and Comparative Social Policy,* 32 (2): 79–82, DOI: 10.1080/21699763.2016.1200112.

Dubbeling, M., H. de Zeeuw, and R. van Veenhuizen. 2011. *Cities, Poverty and Food: Multi-Stakeholder Policy and Planning in Urban Agriculture.* Warwickshire: Practical Action Publishing.

Dubbeling, M., J. Carey, and K. Hochberg. 2016. *The Role of Private Sector in City Region Food Systems.* AK, Leusden, The Netherlands: RUAF Foundation.

Dubbeling, M., G. Santini, H. Renting, M. Taguchi, L. Lançon, J. Zuluaga, L. de Paoli, A. Rodriguez, and V. Andino. 2017. "Assessing and Planning Sustainable City Region Food Systems: Insights from two Latin American Cities". *Sustainability* 9 (8): 1455.

FAO (Food and Agriculture Organization). 2006. *Policy Brief: Food Security.* Accessed 22 November 2018. www.fao.org/fileadmin/templates/faoitaly/documents/pdf/pdf_Food_Security_Cocept_Note.pdf.

FAO (Food and Agriculture Organization). 2011. *Global Food Losses and Food Waste – Extent, Causes and Prevention.* Rome. Accessed 29 November 2018. www.fao.org/docrep/014/mb060e/mb060e00.pdf.

FAO (Food and Agriculture Organization). 2016. *Aquastat.* Accessed 20 November 2018. www.fao.org/nr/water/aquastat/water_use/index.stm.

FAO (Food and Agriculture Organization). 2017. "FAO Expert Meeting for SDG Indicator 2.4". FAO. Accessed 20 November 2018. www.fao.org/sustainability/news/expert-meeting-sdg-indicator/en/.

FAO (Food and Agriculture Organization). 2018. "Climate Change and Your Food: Ten Facts". FAO. Accessed 20 November 2018. www.fao.org/news/story/en/item/356770/icode/.

FAO, IFAD, UNICEF, WFP, and WHO. 2017. *The State of Food Security and Nutrition in the World 2017: Building Resilience for Peace and Food Security*. Rome, FAO.

Fehling, M., B. D. Nelson, and S. Venkatapuram. 2013. "Limitations of the Millennium Development Goals: A Literature Review". *Global Public Health* 8 (10): 1109–1122.

Forster, T., G. Santini, D. Edwards, K. Flanagan, and M. Taguchi. 2015. "Strengthening Urban Rural Linkages Through City Region Food Systems". Paper for a joint UNCRD/UN Habitat issue of *Regional Development Dialogue* 35.

Gilbert, N. 2012. "One-third of Our Greenhouse Gas Emissions Come from Agriculture". *Nature*, 31 October. Accessed 20 November 2018. www.nature.com/news/one-third-of-our-greenhouse-gas-emissions-come-from-agriculture-1.11708.

Griggs, D., M. Stafford-Smith, O. Gaffney, J. Rockström, M. C. Öhman, P. Shyamsundar, W. Steffen, G. Glaser, N. Kanie, and I. Noble. 2013. "Sustainable Development Goals for People and Planet". *Nature* 495 (7441): 305–307.

Holt-Giménez, E., and M. A. Altieri. 2013. "Agroecology, Food Sovereignty, and the New Green Revolution". *Agroecology and Sustainable Food Systems* 37 (1): 90–102.

Horlings, L. G., and T. K. Marsden. 2011. "Towards the Real Green Revolution? Exploring the Conceptual Dimensions of a New Ecological Modernisation of Agriculture that Could 'Feed the World'". *Global Environmental Change* 21 (2): 441–452.

Kloppenburg Jr, J., S. Lezberg, K. De Master, G. Stevenson, and J. Hendrickson. 2000. "Tasting Food, Tasting Sustainability: Defining the Attributes of an Alternative Food System with Competent, Ordinary People". *Human Organization* 59 (2): 177–186.

Mathers, C. D., and D. Loncar. 2006. "Projections of Global Mortality and Burden of Disease from 2002 to 2030". *PLoS Medicine* 3 (11): e442.

McIntyre, B. D. 2009. *International Assessment of Agricultural Knowledge, Science and Technology for Development (IAASTD): Global Report*. McGraw-Hill Publishing.

McMichael, P. 2011. "Food System Sustainability: Questions of Environmental Governance in the New World (Dis)order". *Global Environmental Change* 21 (3): 804–812.

Moragues-Faus, A., and T. Marsden. 2017. "The Political Ecology of Food: Carving 'Spaces of Possibility' in a New Research Agenda". *Journal of Rural Studies* 55: 275–288.

Morley, A., and T. Marsden. 2014. "Current Food Questions and Their Scholarly Challenges: Creating and Framing a Sustainable Food Paradigm". In *Sustainable Food Systems*, edited by T. Marsden and A. Morley, 1–29. London: Routledge.

Pimbert, M. ed., 2017. *Food Sovereignty, Agroecology and Biocultural Diversity: Constructing and Contesting Knowledge*. Abingdon: Routledge.

Pogge, T., and M. Sengupta, M. 2016. "Assessing the Sustainable Development Goals from a Human Rights Perspective". *Journal of International and Comparative Social Policy* 32 (2): 83–97.

Prain, G., and M. Dubbeling. 2011. *Urban Agriculture: A Sustainable Solution to Alleviating Urban Poverty, Addressing the Food Crisis, and Adapting to Climate Change*. Synthesis Report on Five Case Studies Prepared for the World Bank. Leusden: RUAF Foundation.

Schuyt, K., and L. Brander. 2004. "Living Waters Conserving the Source of Life: The Economic Values of the World's Wetlands". In *Living Waters Conserving the Source of Life: The Economic Values of the World's Wetlands*. WWF.

Schwendler, S. F., and L. A. Thompson. 2017. "An Education in Gender and Agro-ecology in Brazil's Landless Rural Workers' Movement". *Gender and Education* 29 (1): 100–114.

Seufert, V., N. Ramankutty, and J. A. Foley. 2012. "Comparing the Yields of Organic and Conventional Agriculture". *Nature* 485 (7397): 229–232.

Seyfang, G. 2006. "Ecological Citizenship and Sustainable Consumption: Examining Local Organic Food Networks". *Journal of Rural Studies* 22 (4): 383–395.

Sonnino, R., and T. Marsden. 2005. "Beyond the Divide: Rethinking Relationships between Alternative and Conventional Food Networks in Europe". *Journal of Economic Geography* 6 (2): 181–199.

Sustainable Development Knowledge Platform. 2018. "Voluntary National Reviews Database". *Sustainable Development Knowledge Platform*. Accessed 26 November 2018. https://sustainabledevelopment.un.org/vnrs/.

Taylor, M. 2018. "Climate-Smart Agriculture: What is it Good For?" *The Journal of Peasant Studies* 45 (1): 89–107.

UNDESA (United Nations Department of Economic and Social Affairs). 2017. *Voluntary National Reviews: Synthesis Report*. Accessed 26 November 2018. https://sustainable development.un.org/content/documents/17109Synthesis_Report_VNRs_2017.pdf.

UNESCO (United Nations Educational, Scientific and Cultural Organization). 2012. "World Water Assessment Program". Accessed 20 November 2018. www.unesco.org/new/en/natural-sciences/environment/water/wwap/wwdr/wwdr4-2012/.

UNESCO (United Nations Educational, Scientific and Cultural Organization). 2014. *World Water Development Report 2014: Water and Energy*. Vol. 1. Accessed 20 November 2018. http://unesdoc.unesco.org/images/0022/002257/225741E.pdf.

UNGA (United Nations General Assembly). 2015. *Transforming our World: The 2030 Agenda for Sustainable Development*. A/RES/70/1, October 21. Accessed 16 October 2018. https://undocs.org/A/RES/70/1.

United Nations. 2014. "Water, Food and Energy". *UN Water*. Accessed 20 November 2018. www.unwater.org/water-facts/water-food-and-energy/.

United Nations. 2017. "Sustainable Development: Goal 2". *United Nations*. Accessed 30 November 2018. https://sustainabledevelopment.un.org/sdg2.

UN-Sustainable Development Goals. 2017. "Sustainable Development Knowledge Platform: Voluntary National Reviews Database". *United Nations*. Accessed 26 November 2018. https://sustainabledevelopment.un.org/vnrs/.

WHO (World Health Organization). 2017a. "Obesity and Overweight Fact Sheet". *WHO*, 16 February. Accessed 20 November 2018. www.who.int/mediacentre/factsheets/fs311/en/.

WHO (World Health Organization). 2017b. "Diabetes: Key Facts". *WHO*, 30 October. Accessed 20 November 2018. www.who.int/mediacentre/factsheets/fs312/en/.

Wichterich, C. 2015. "Contesting Green Growth, Connecting Care, Commons and Enough". In *Practicing Feminist Political Ecologies: Moving beyond the "Green Economy"*, edited by W. Harcourt, and I. L. Nelson, 67–100. London: Zed Books.

3 From MDGs to SDGs

Health slips in global priorities

Alan Whiteside

Introduction

The establishment of the World Health Organization (WHO) in 1946 marked the acknowledgement of health as an international issue. The definition of health adopted by WHO was "a state of complete physical, mental and social well-being and not merely the absence of disease or infirmity" (WHO 2006). Thirty-two years later, in 1978, the Declaration of Alma-Ata was adopted at the International Conference on Primary Health Care (PHC) at Almaty (formerly Alma-Ata), Kazakhstan. This set out goals for global health and was, effectively, a revised mission statement for the WHO. It expressed the need to protect and promote the health of all people, underlining the importance of primary health care. In 2018, the 40th anniversary of the declaration, there was much analysis of its importance and achievements. The current global emphasis is on Universal Health Coverage (UHC).

The World Bank's 16th Annual Report in 1993 with the theme "Investing in Health" acknowledged the interplay between human health, health policy and economic development (World Bank 1993). It raised the profile and importance of health at a time when there was optimism about global health, a sense that infectious disease could conquered. This was premature, the unexpected and catastrophic AIDS epidemic was emerging and was to drive thinking around health and its governance.

In 2000 the eight Millennium Development Goals (MDGs) were adopted following the Millennium Summit and Millennium Declaration of the United Nations. The sixth goal was "To combat HIV/AIDS, malaria, and other diseases".

Negotiations for the replacement for the MDGs began in January 2015 and ended with the final text being adopted at the UN Sustainable Development Summit in New York, in September 2015. The 193 countries of the UN General Assembly approved the Agenda: "Transforming our World: the 2030 Agenda for Sustainable Development" (UNGA 2015). This has 92 paragraphs. Paragraph 51 outlines the 17 Sustainable Development Goals (SDGs) and the associated 169 targets. The SDGs have as Goal 3 (Good Health and Well-Being) with 13 specific targets.

This chapter provides a brief history of the evolution of public health, and its challenges and advances. It relates these to international efforts, arguing that health received greater priority in the MDGs, but that growing concerns about the social determinants of health, as well as greater confidence that the HIV and AIDS epidemic is being addressed, may explain the shift in priorities in the SDGs.

A short history of public health

The focus, when looking at health and the SDGs, is on public health rather than treatment of individuals. Early "public health" focused on preventing the spread of disease, mainly by isolating those who were ill through quarantine. In the late seventeenth century advances in public health began. As is eloquently argued by Davies *et al.* (2014), there were five waves of public health development. The first (1830 to 1900) was structural: addressing physical and environmental conditions, ensuring clean water and sewerage disposal, legislating for food safety, and improved working conditions. The action on sanitation was spurred by the 1858 "Great Stink" of London.

The second wave (1890 to 1950) was characterized by biomedical and scientific advances. These built on the successful experimental vaccination against smallpox in 1796. The United Kingdom's 1853 Vaccination Act made immunization for smallpox compulsory for infants by the age of three months. The idea of sterilization and hygiene began with Ignaz Semmelweis, a Hungarian doctor in the Vienna General Hospital. He observed in 1846 that women giving birth in the medical student/doctor-run maternity ward were more likely to die than women in the midwife-run ward. Doctors and students often visited the maternity ward directly after performing autopsies. Midwives did not conduct surgery or autopsies, so were not carrying infections. Compulsory hand washing had an immediate impact on morbidity and mortality.

The third wave (1940 to 1980) was clinical and scientific, with growing understanding of causes of communicable disease, and development of treatments. Alexander Fleming identified the first antibacterial medicine, penicillin, in 1928. Rapid advances in cancer treatment included drugs, radiation, and surgery.

The fourth wave began in 1960 with examination of social determinants of health and understanding that much ill health has roots in economic and social factors. In 2005 the WHO established the Commission on Social Determinants of Health (CSDH) to address the social factors leading to ill health and health inequities (Commission on the Social Determinants of Health 2008).

Davies *et al.* (2014) argue that the fifth wave will be "a culture for health" where health should be the norm: healthy choices valued and incentivized and unhealthy activities vigorously discouraged. This might have application in the rich world; it is less clear how people in developing nations will be able to respond. A normative culture for health would have far-reaching implications for global health and fits with the SDGs.

Concerns about well-being, a more complicated idea than physical health, are gaining ground. Well-being is about both physical and mental health and it is mental health that has been most neglected. The WHO's definition of mental health is "a state of well-being in which the individual realizes his or her own abilities, can cope with the normal stresses of life, can work productively and fruitfully, and is able to make a contribution to his or her community" (WHO 2005). The Public Health Agency of Canada's definition of a positive state of mental health is:

> The capacity of each of us to feel, think, and act in ways that enhance our ability to enjoy life and deal with the challenges we face. It is a positive sense of emotional and spiritual well-being that respects the importance of culture, equity, social justice, interconnections and personal dignity
> (Canadian Institute for Health Information 2009)

Mental ill health is a growing global problem. In 2017 the UK's The Mental Health Foundation reported 40 per cent of people had experienced depression and 25 per cent had panic attacks. This is differentiated by age, gender and income. People with lower incomes or who are unemployed; younger people; and women are more likely to face mental health challenges (The Mental Health Foundation 2017). According to the Institute of Health Metrics and Evaluation (IHME) mental disorders were the sixth most important cause of loss of disability adjusted life years (DALYs) globally in 2016. This has increased everywhere, in all income categories, since 1990 (IHME 2018a).

Public health challenges and advances: 1945 to 2018

Health in development

Because good health increases productivity of individuals and hence economic growth, investing in health should mean more and faster development (Bloom, Khoury, and Subbaraman 2018). The World Bank's 1993 World Development Report (WDR) (World Bank 1993, iii) examined "the interplay between human health, health policy and economic development".

The 1993 WDR was based on a single paper commissioned from "The Global Burden of Disease Study" (GBD) which began in 1990. The original project quantified the health effects of more than 100 diseases and injuries for eight world regions. It gave estimates of morbidity and mortality by age, sex, and region, and introduced the concept of DALYs to compute the burden of diseases, injuries, and risks. DALYs are defined as the sum of years of potential life lost due to premature mortality and years of productive life lost due to disability.

Subsequent work was led by the WHO, with its GBD studies being produced up to 2007. In July of that year the IHME was launched as an independent global health research centre at the University of Washington. Its goal was to provide an impartial, evidence-based picture of global health trends to inform the work of policymakers, researchers, and funders. It has largely taken over the function of producing these data from other health institutes, including the WHO. The main support to the IHME has been from the Bill & Melinda Gates

Foundation and the US state of Washington. The output, while not without problems and controversy, is an invaluable public good in understanding of, and policymaking in, health (Shiffman 2014; Grépin 2015; Kaasch 2015).

The World Bank 1993 report noted that life expectancy especially in the developing world had risen and child mortality had decreased. The burden of childhood and tropical diseases was high but there were new challenges. Specifically identified were "AIDS and the diseases of aging populations", the latter being Non-Communicable Diseases (NCDs). Although health was on the agenda so were the challenges: all countries were struggling with the problems of controlling health expenditures and making health care accessible.

The WDR's findings based on the GBD research included estimates of the burden and the cost-effectiveness of interventions. These were specifically to help set priorities for health spending for governments in the developing world and the former socialist countries (World Bank 1993). The key interventions were: economic policies to empower households; moving government health spending from tertiary to primary care (including prevention); and encouraging diversity and competition in health provision. This was very much a report of its time; it did not focus on the problems of the poorer countries. It did, however, identify the emerging AIDS epidemic as a looming challenge.

Measuring global disease burden

The GBD provides an extensive interactive data base (IHME 2018a). This illustrates the differences in the disease burdens in the rich and poor worlds, and since data is available from 1990 it shows the trends. The data in Tables 3.1, 3.2, and 3.3 are for DALYs; Table 3.1 for Global DALYs, Table 3.2 for High

Table 3.1 Global disease burden ranking, both sexes, all ages, DALYs per 100,000

1990 rank	2016 rank
1 Diarrhea/LRI/other ID	1 Cardiovascular disorders NCD
2 Cardiovascular disorders NCD	2 Diarrhea/LRI/other ID
3 Neonatal disorders ID	3 Neoplasms NCD
4 Other non-communicable NCD	4 Other non-communicable NCD
5 Neoplasms NCD	5 Neonatal disorders ID
6 Unintentional injury I	6 Mental disorders NCD
7 Mental disorders NCD	7 Musculoskeletal disorders NCD
8 NTDs and malaria ID	8 Diabetes/urog/blood/endocrine NCD
9 Chronic respiratory disorders NCD	9 Unintentional injury I
10 Musculoskeletal disorders NCD	10 Neurological disorders NCD

Source: Table derived from the Global Burden of Disease interactive website. https://vizhub.health data.org/gbd-compare/ Accessed 28 November 2018.

Notes
NTDs refer to neglected tropical diseases; urog refer to urinogenitary diseases; LRI refers to lower respiratory infections. ID refers to infectious diseases, NCD refers to Non-Communicable Diseases and I refers to injuries.

Table 3.2 Global disease burden ranking, high SDI countries, both sexes, all ages, DALYs per 100,000

1990 rank	2016 rank
1 Cardiovascular disorders NCD	1 Neoplasms NCD
2 Neoplasms NCD	2 Cardiovascular disorders NCD
3 Mental disorders NCD	3 Musculoskeletal disorders NCD
4 Other non-communicable NCD	4 Mental disorders NCD
5 Musculoskeletal disorders NCD	5 Other non-communicable disorders NCD
6 Neurological disorders NCD	6 Neurological disorders NCD
7 Diabetes/urog/blood/endocrine NCD	7 Diabetes/urog/blood/endocrines NCD
8 Unintentional injury I	8 Unintentional injury I
9 Transport injuries I	9 Chronic respiratory disorders NCD
10 Chronic respiratory disorders NCD	10 Self-harm and violence I

Source: Table derived from the Global Burden of Disease interactive website. https://vizhub.health data.org/gbd-compare/ Accessed 28 November 2018.

Table 3.3 Global disease burden ranking, low SDI countries, both sexes, all ages, DALYs per 100,000

1990 rank	2016 rank
1 Diarrhea/LRI/other ID	1 Diarrhea/LRI/other ID
2 Neonatal disorders ID	2 Neonatal disorders ID
3 NTDs & malaria ID	3 NTDs and malaria ID
4 Nutritional deficiencies ID	4 HIV/AIDS and Tuberculosis ID
5 HIV/AIDS and Tuberculosis ID	5 Cardiovascular disorders NCD
6 Other non-communicable disorders NCD	6 Other non-communicable disorders NCD
7 Cardiovascular disorders NCD	7 Nutritional deficiencies ID
8 Unintentional injuries I	8 Unintentional injury I
9 Other group 1 (Communicable, maternal, neonatal and nutritional diseases) I	9 Mental disorders NCD
10 Diabetes/urog/blood/endocrine NCD	10 Neoplasms NCD

Source; These tables are derived from the Global Burden of Disease interactive website. https:// vizhub.healthdata.org/gbd-compare/ The setting used were DALYs from all causes by location (High SDI and Low SDI), comparing 1990 to 2016 for all ages and both genders.

Socio-Demographic Index (SDI) Countries, and Table 3.3 for Low SDI countries.

The SDI is a new metric, which first appeared in the GBD update published in October 2016. It is becoming an integral part of the GBD project. The SDI is a "summary measure of a geography's socio-demographic development based on average income per person, educational attainment, and total fertility rate (TFR)" (IHME 2018b). The data can be "sliced and diced" in many ways using the interactive data visualization tools. There are five categories of nations: high SDI, high-middle SDI, middle SDI, low-middle SDI, and low SDI. The tables display how the global disease burden ranking has evolved from 1990 to 2016, and how this differs for the highest and lowest SDI countries.

Globally the top cause of DALYs was diarrhoeal diseases in 1990, followed by cardiovascular disease; in 2016 these reversed, with cardiovascular disease first and diarrhoeal disease second. Cancers moved up to third place in global DALYs. In low SDI countries the top four causes of lost DALYs are communicable diseases followed by cardiovascular disease and other NCDs. In high SDI countries the top seven causes of lost DALYs are NCDs. If the data are interrogated at the country level then in Southern and Eastern Africa HIV, AIDS and tuberculosis (TB) top the tables. For example, in Eswatini (Swaziland), in 1990 HIV, AIDS and TB accounted for 2,798 DALYs per 100,000 people, by 2016 it was 20,718 per 100,000.

The AIDS epidemic and other communicable diseases

High rates of morbidity and mortality due to infectious disease remain a challenge in developing countries. The leading contributors to the disease burden are the "Big Four": HIV and AIDS, malaria, TB, and hepatitis C virus (HCV). The links between HIV and TB are well documented. TB is one of the main opportunistic infections HIV positive people are likely to experience, especially in poor countries. Investment and efforts to combat infectious diseases have borne fruit, but Africa remains most affected. With almost all disease (including non-communicable ones) prevention is better, and generally cheaper, than cure. While currently we may not definitively know the causes of many cancers, life style and environmental factors are implicated in their development. With increased knowledge these too may be (more) preventable.

The progress in addressing communicable diseases was partly driven by experience of, and response to, the HIV and AIDS epidemic. This new disease was first identified in the USA in 1981 with cases soon being recorded across the world. Within a few years a large number of cases were reported from central and eastern Africa, which was to become the centre of the epidemic.

The name "Acquired Immunodeficiency Syndrome" (AIDS) was agreed in Washington in July 1982. In 1983 the virus was identified by the Pasteur Institute in France and called "Lymphadenopathy-Associated Virus" or LAV. In 1987, the name "Human Immunodeficiency Virus" (HIV) was confirmed by the International Committee on Taxonomy of Viruses. It was not until 1996 that effective treatment was developed. Early drugs were expensive and beyond the reach of the worst affected countries until the early 2000s.

AIDS was a stark reminder of the spread of diseases from animals to humans (zoonosis). Recent examples include Severe Acute Respiratory Syndrome (SARS), tracked to civet cats; Avian Influenza (bird flu); Middle East Respiratory Syndrome (MERS), linked to camels; the Ebola virus carried by fruit bats; and the Zika Virus, which was first reported in monkeys in Uganda in 1947, but which is now spread by mosquitoes. HIV is, so far, the deadliest pathogen to have made the leap across the species barrier to humans. SARS was, fortunately, not as infectious; avian flu has not (yet) taken hold in humans; MERS outbreaks have been infrequent and controlled. The Ebola outbreak requires constant monitoring but has been contained. It seems Zika will be managed.

The frequency with which zoonotic diseases are emerging is increasing. This is, in part, due to human activity both direct and indirect. Examples of the former come from virgin forest being cleared resulting in more human: host: pathogen contact. AIDS came from chimpanzees in central Africa and sooty mangabey monkeys in West Africa. Further there can be no doubt that human activities are leading to environmental change (see the chapter by Dalby, Chapter 8). In recent times, apart from AIDS, humankind has dodged the bullet of an uncontrolled pandemic.

The health governance for new and emerging diseases is reasonably well developed at both the national and international levels. The United States' agency, the Centres for Disease Control, based in Atlanta has had, to date at least, a global presence. It has monitored outbreaks and helped co-ordinate responses. The WHO tracks evolving infectious diseases and will "sound the alarm when needed, share expertise, and mount the kind of response needed to protect populations from the consequences of epidemics, whatever and wherever might be their origin" (WHO 2018a). Following the Ebola outbreak in 2014 the World Bank launched the Pandemic Emergency Financing Facility (PEF), in May 2016. This is intended to be an "innovative, fast-disbursing global financing mechanism designed to protect the world against deadly pandemics" (World Bank 2016). Launching this Jim Yong Kim, the President of the Bank, said:

> Pandemics pose some of the biggest threats in the world to people's lives and to economies, and for the first time we will have a system that can move funding and teams of experts to the sites of outbreaks before they spin out of control.
>
> (World Bank 2016)

Three major governance challenges for global health were identified by Frenk and Moon (2013). The first is sovereignty: health is a national (or even a state or provincial) responsibility; however, the increasingly globalized nature of the world means health issues need co-operative responses. Second, health is not a bounded sector; it is linked with other sectors – for example, trade and the movement of people – both as consumers (patients) and providers (health care workers). Finally, who is responsible for ensuring health, and who can hold others accountable? The WHO's World Health Assembly is the closest to international representation but is under-resourced and has limited ability to enforce decisions.

AIDS accelerated changes in global governance of health and speaks to the challenges identified above. There were early indications of its potential political, economic, and social impact. In January 2000 during the United States' rotating presidency of the UN Security Council, Richard Holbrooke, the US Ambassador to the UN, convened a session to discuss AIDS in Africa, presided over by US Vice President Al Gore. This marked the first time a health issue was discussed as a threat to peace and security. AIDS was declared to be a security threat to all nations (United Nations Security Council 2000).

When the MDGs were adopted in 2000, HIV was still sweeping across the world. Although it was generally accepted it would not be a major, out-of-control problem in the developed world, it was not clear how far it would spread across the developing countries. In particular, in parts of Africa prospects seemed bleak as the numbers of infected people, and the death toll, continued to rise. Treatment remained expensive. This was the context for the HIV MDG.

It is certainly true that communicable diseases are best prevented, and wealthy countries have an interest in containing outbreaks, preferably outside their borders. This harks back to the biblical quarantine of lepers and the twelfth century Venetian *quaranta giorni*, which in the local dialect meant "forty days", an enforced period of isolation of ships and people, before they entered the city-state of Ragusa (modern Dubrovnik, Croatia). The impact of disease was seen with the 2003 Canadian SARS outbreak. This was concentrated in Toronto and infected only 257 individuals in the province of Ontario, but it caused widespread fear. Once a disease reaches a wealthy country costs go up, panic ensues, and citizens lose faith in governments.

Non-communicable diseases (NCDs)

An NCD is a medical condition or disease not caused by infectious agents and not transmissible. NCDs are diseases linked to aging (e.g. musculoskeletal disease), lifestyle (e.g. diabetes), and environment (e.g. transport injuries). Some NCDs are chronic, last a long time and progress slowly; others, such as heart disease or cancer, can develop rapidly. There are those that detract from quality of life and require treatment for life (glaucoma, for example), while others, such as cataracts, can be treated and reversed. Risk factors such as a person's background, lifestyle, and environment increase the likelihood of some NCDs: about five million people die from tobacco use; 2.8 million die from obesity; roughly 2.6 million deaths result from high cholesterol; and 7.5 million from hypertension.

Most cancers are triggered by environmental or lifestyle-related risk factors such as using tobacco or excessive alcohol consumption, being obese, not taking exercise, poor diet, sexually transmitted infections, and pollution – especially air pollution. Cardiovascular disease is a major killer and is linked to lifestyle. Type 2 diabetes, increasingly common in the developing world is preventable and manageable, and it can even be reversed with lifestyle changes. Mental disorders do not often result in death but rank fourth in the 2016 DALY tables for high SDI countries and ninth for the low SDI countries. These are growing in importance especially among adolescents and young adults.

The UN General Assembly convened the first High-Level Meeting on NCDs on 19–20 September 2011. The background papers noted "The four main non communicable diseases – cardiovascular disease, cancer, chronic lung diseases and diabetes – kill three in five people worldwide" and "cause great socio-economic harm within all countries, particularly developing nations" (UNGA 2011). In September 2018, the third High-Level Meeting on the prevention and control of NCDs was convened by the UN. This undertook a comprehensive review of global and national progress (WHO 2018d).

The major problem with NCDs and national and international health governance is around the attention they command. In terms of Davies *et al.*'s (2014) waves of public health, we are in the fifth wave: the culture for health. Here health should be the norm, with healthy choices valued and incentivized. Because NCDs can be seen as self-inflicted (lung cancer is caused by smoking, diabetes by diet and obesity), are chronic requiring long-term treatment, and are generally incurable there is not the appetite in the international development community for attention to these issues. AIDS, of course requires huge amounts of money in the long term, and its acquisition has moral implications for some. It is the health exception and as such is deeply interesting to track how it has driven agendas and the discourse, and what the future might hold.

The international agenda develops to include health

In September 2000, world leaders came together at UN Headquarters in New York to adopt the United Nations' Millennium Declaration. This committed the countries to a global partnership to reduce extreme poverty. It set a series of time-bound objectives, the eight MDGs, to run from 2000 to 2015. Each had specific targets, and dates for achieving them.

All goals were linked to the creation of healthier populations. Eradicating extreme poverty and hunger, MDG 1, aimed to reduce the proportion of people living on less than $1.25 a day; achieve decent employment; and halve the proportion of people who suffer from hunger. The importance of education in creating healthy populations has been well documented so MDG 2 (Achieving universal primary education) would have significant health benefits. The same was true of MDG 3: (Promote gender equality and empower women).

The main health Goals were 4, 5, and 6. MDG 4 was to reduce child mortality rates, specifically to reduce the under-five mortality rate by two-thirds. MDG 5 was to improve maternal health and reduce the maternal mortality ratio by three-quarters, increase the proportion of births attended by skilled health personnel, and achieve universal access to reproductive health. MDG 6 was to combat HIV/AIDS, malaria, and other diseases. The targets were to halt and begin to reverse the spread of HIV/AIDS by 2015; to achieve universal access to treatment for HIV/AIDS for all who needed it by 2010, and to halt and begin to reverse incidence of malaria and other major diseases by 2015. The final two goals, MDG 7 (Ensure environmental sustainability) and MDG 8 (Develop a global partnership for development), are important for improving health, but more tangentially. However, Target 8.e was specific: "In co-operation with pharmaceutical companies, provide access to affordable, essential drugs in developing countries".

The MDGs must be seen in context: they were a unique set of globally agreed, simple, and measurable targets. The prominence of health in the goals was exceptional. Its inclusion was largely due to AIDS; the uncertainty as to how far and fast the epidemic would spread especially in countries targeted by the goals; and the fact that treatment was in its infancy and unaffordable in the developing world. AIDS activists and advocates from around the world were vocal and convincing. The Joint United Nations Programme on HIV/AIDS (UNAIDS) had been in existence for four years and was in an expansionary phase.

The MDGs were ambitious and not fully achieved; however, there is consensus that they were successful in focusing the world's attention on key targets (Galatsidas and Sheehy 2015; IHME 2018c). Of course, identifying cause and effect is complex and it is possible that the gains would have occurred anyway (McArthur and Rasmussen 2017). In the development continuum the MDGs provided a solid foundation for the SDGs.

The Sustainable Development Goals and health

As the end of the MDGs approached in 2015, the United Nations and the global community began an extensive and lengthy consultation process, involving the 193 member states and global civil society, which resulted in the 17 SDGs, and 169 targets (UNGA 2015). The resolution is a broad intergovernmental agreement that "acts as the Post-2015 Development Agenda" (UNGA 2015). As is discussed health is no longer as much of a priority, and this "demotion" can perhaps be linked to the health advances due to the AIDS epidemic and the (wrong) perception that it is under control. The SDGs were agreed at a different time – post 9/11 and post 2008 crash and were more bottom-up.

As with the MDGs, meeting any, or all, of the SDGs would improve health. However, the specific health goal is number three: Good Health and Well-Being for People to "Ensure healthy lives and promote well-being for all at all ages" (UNGA 2015, 3). There are nine numbered "uber" targets and four with "sub" targets, all to be met by 2030; it is both a shopping and a wish list.

The health challenges: are the SDGs appropriate?

The majority of the SDGs are not specifically related to health, but all will influence it by creating healthier societies and environments. For example, quality education improves child development and increases opportunities for young people. SDG 6 (providing clean water and sanitation) is a basic public health response. If this were achieved, it would ensure a decrease in waterborne diseases, a major killer in the developing world. Improved nutrition will improve health, with nutrition in turn depending on agricultural production and dietary diversity, discussed by Blay-Palmer and Young (Chapter 2).

The emphasis in the SDGs in moving away from communicable diseases and towards the non-communicable disease is driven by the increasing burden of

NCDs. The third High-Level Meeting on NCDs in September 2018 flowed from and supported the SDGs. The goal of reducing premature mortality from NCDs by one-third by 2030 was adopted and a global progress monitor and country by country scorecards were developed. It is hoped NCDs will be integrated into national SDGs to ensure the most effective interventions are identified and implemented, especially with regard to the Universal Health Coverage packages; and there will be agreement on increased taxes on tobacco, alcohol, and sugar.

Unfortunately, the SDGs do not focus on some of the key challenges to improving health. The first is health care delivery and policy. Health policy is developed nationally but funding is allocated to and spent by provincial or local departments of health and there are variations in need and delivery (Whiteside 2014). The cost of out of pocket care pushes about 100 million people into extreme poverty yearly (Angenendt and Voss 2018).

The second area is the collision of epidemics, in particular HIV and TB and NCDs. This situation is most pronounced in Southern Africa and best recorded in South Africa as the Medical Research Council has a Burden of Disease Research Unit (Bradshaw *et al.* 2016) and Parliament takes an active interest in these issues (NCOP Social Services 2016). All too common across the developing world are under and over nutrition, obesity, alcohol consumption, and interpersonal violence.

The third area concerns migrants and refugees (see also the chapters by Crush, Chapter 6, and KC and Hennebry, Chapter 5). The migrants across the Mediterranean and the Mexican border and the Rohingya refugee crisis are well publicized. In 2017, there were 68.5 million displaced people of whom 25.4 million were refugees and three million were asylum seekers. These people need health care both during periods of transition and in the countries in which they end up. The process of moving is hazardous, increasing risks of violence and injuries, as well as threatening consistency of treatment for NCDs. Infectious disease outbreaks may occur in conflicts and humanitarian crises and are harder to treat (Angenendt and Voss 2018).

Fourth, the SDGs recognize the challenges of climate change but not how they interact with health. Droughts will increase in frequency and magnitude, with consequences for production of and access to food. The quality of the air people breathe and the heat in the environments in which they live are changing and not for the better. Cape Town was close to becoming the first major city to run out of drinking water. In mid-January 2018 water was projected to run out on 22 April 2018 (Morrison 2018). The city capped household water usage at 87 litres (23 gallons) per person, per day. Fortunately, the winter rains in 2018 brought relief.

Fifth, there are issues of funding. The SDGs come with a price tag, but international resources are diminishing. This is particularly marked in HIV and AIDS. According to a CSIS analysis, donor commitments fell by $3 billion between 2012 and 2017 and eight of the 14 major donors decreased their funding in 2017 (Morrison 2018). The Trump administration proposed a 30 per cent decrease in the US foreign affairs budget that would have had a dramatic effect on global health programs, if implemented. While the risk of this happening has receded, the threat remains. Horton (Chapter 13) discusses how the much larger resource needs of the SDGs can be met, with specific examples for health.

Finally, there has been commentary as to whether the SDGs are fit for purpose. Frenk and Moon (2013) remind readers that "the organised social response to health conditions at the global level is what we call the global health system, and the way it is managed is what we refer to as governance". There is no accountable system of global governance and certainly none for health. The WHO's 2007 International Health Regulations (IHR) are supposed to be an international legally binding instrument that helps prevent and respond to immediate public health risks "that have the potential to cross borders and threaten people worldwide [and which] … require countries to report certain disease outbreaks and public health events to WHO" (WHO 2018b). However, there is no way to enforce these.

A more puissant, and recent, critique by van de Pas *et al.* (2017) asks if global health governance in the SDG is grounded in the "right to health". The authors query if "the existing human rights legislation would enable the principles and basis for the global governance of health beyond the premise of the state". They suggest there is a governance gap between the human rights framework and what actually goes on in global health and development policies. "The central question … was: does the SDG agenda overcome that gap? Does the SDG agenda entail new or improved global health governance that satisfies the demands of the Right to Health? The answer is, unfortunately, negative". The SDG health agenda, they say, undercuts the Right to Health.

Limits to health

This chapter notes the relative lack of direct health targets in the SDGs. There is though the implicit desire to make the world a better and healthier place. The SDGs speak to the social determinants of health which in turn influence individual and group health. SDGs are health-promoting through better living and working conditions (such as distribution of income, wealth, influence, and power), rather than directly on individual risk factors (such as behavioural risk factors or genetics) that influence the threat of a disease, or vulnerability to disease or injury.

The WHO's Commission on Social Determinants of Health 2008 report "Closing the Gap in a Generation" identified two broad areas of social determinants of health. First were daily living conditions, including healthy physical environments, fair employment and decent work, social protection across the lifespan, and access to health care. The second was distribution of power, money, and resources within and between societies.

The 2011 World Conference on Social Determinants of Health brought 125 member states together and resulted in the Rio Political Declaration on Social

Determinants of Health. This declaration affirmed that health inequities are unacceptable and noted that these "inequities arise from the societal conditions in which people are born, grow, live, work, and age, ... effective prevention and treatment of health problems" (WHO 2018c, 2).

If social determinants are seen as crucial, development has to be at a broader level than health. This has been supported by numerous studies. A 2017 meta-analysis in *The Lancet* looked at individual-level data from 48 independent prospective cohort studies. There was information about socio-economic status, indexed by occupational position, risk factors and mortality, for 1.75 million individuals in seven high-income countries. Unsurprisingly it found people with low status had greater mortality than those with high status.

> Low socioeconomic status was associated with a 2.1-year reduction in life expectancy between ages 40 and 85 years, the corresponding years-of-life-lost were 0.5 years for high alcohol intake, 0.7 years for obesity, 3.9 years for diabetes, 1.6 years for hypertension, 2.4 years for physical inactivity, and 4.8 years for current smoking.
>
> (Stringhini *et al.* 2017)

Social determinants are shaped by politics and public policies that reflect the influence of prevailing political ideologies. Apart from setting standards and norms there is little international players can do to improve the social determinants without buy-in from developing country governments. Indeed, the global trend seems that the well-off and healthy become even richer and the poor, who are already more likely to be ill, become even poorer. This is sometimes known as the Matthew effect: "For unto everyone that hath shall be given, and he shall have abundance: but from him that hath not shall be taken away even that which he hath".[1]

Addressing the social determinants of health can undoubtedly extend lives and reduce ill-health during those lives. The question of how to do this cost-effectively is one that vexes policymakers and economists. The 2018 mantra is universal health coverage (UHC). This is a system for providing basic health care and financial protection to all. It does not propose that everyone would get everything; there may be variation on who is covered, and what services and how much of the cost is covered. Ultimately the WHO sees it as way in which citizens can get health services without financial hardship.

In his critique of UHC, Matheson (2015) notes that Margaret Chan, then-WHO head, described UHC as the "single most powerful concept that public health has to offer" as it unifies "services and delivers them in a comprehensive and integrated way" (Matheson 2015). He concludes:

> It (UHC) has certainly been established as a political issue of high importance in the context of the SDGs, and some courage is being demonstrated from the two leaders (WHO and the World Bank) of the lead agencies. The policy itself is less well developed, and the risks of addressing global

policy primarily through a global monitoring and targeting mechanisms have not been adequately addressed. Supporting social struggle and action is a weakness of the current approach, and the soundness of the evidence base remains a work in progress.

(Matheson 2015)

The concept of UHC is articulated most clearly in targets 3.7 and 3.8. Target 3.7 seeks to ensure universal access to sexual and reproductive health-care services and 3.8 to achieve universal health coverage. These are then the "health" SDGs that advocates and professionals need to highlight.

Is there the governance and global commitment? The concern was raised in a letter to WHO and the global health community from Chancellor Angela Merkel, Ghanaian President Nana Akugo-Addo, and Norwegian Prime Minister Erna Solberg. They called for "greater commitment to the implementation of the 2030 Agenda for Sustainable Development and its sustainable development goals".[2] Angenendt and Voss (2018) suggest three policy options: "Strengthen the WHO financially without making it depend on a few donors.... Create a binding Framework Convention on Global Health, similar to the Framework Convention on Tobacco Control (FCTC); and ... Improve engagement by beginning a systematic dialogue".

The challenges for global health governance have been addressed. Frenk and Moon (2013) identify the four functions for a global health system. The production of global public goods such a surveillance, reporting and setting standards; management of externalities, for example pollution produced in one nation affecting another; mobilization of global solidarity; and stewardship, as achieving health Pareto Efficiency. These are a reflection on limits of health and an agenda for optimizing global health governance.

Conclusion

There have been huge advances in health. Life expectancy has increased (except in some countries in the 2000s when AIDS caused a reversal) and people are living healthier lives. The overall challenge for the global community may not be to extend lives even further but to improve survival of the young, and increase quality of life, including well-being and mental health. The World Bank noted good health should be a goal in itself (World Bank 1993).

The Copenhagen Consensus Center researches and prioritizes solutions to the world's biggest problems from an economic point of view. In their review of "the smartest targets for the post 2015 development agenda" they estimate the social, economic and environmental benefits of a dollar spent on various interventions. Their top five interventions (with the returns per dollar) are: reduce world trade restriction (full Doha) $2,011; freer regional Asian Pacific trade $1,299; universal access to contraception $120; aspirin heart attack therapy $63; and expanded immunization $60 (Copenhagen Consensus Center 2018).

Limited resources mean we have to prioritize. The Copenhagen Consensus "provides information on which targets will do the most social good relative to their costs" (Copenhagen Consensus Center 2018). They identify 76 possible areas of spending, 16 of which are directly health related and have positive rates of return. The final decisions on spending are not decided purely on economics; however, knowing the costs and benefits provides guidance and a framework.

Health professionals have long understood the importance of attitudes and equity in health. This was addressed innovatively and coherently by Wilkinson and Pickett in their 2010 book. Their latest publication takes these ideas to a new level (Wilkinson and Pickett 2018). They focus on how inequality affects individuals and how "when the gap between rich and poor increases, so does the tendency to define and value ourselves and others in terms of superiority and inferiority". This circles to Davies *et al.*'s (2014) fifth wave of public health: the culture for health, which is both individualized and can be a public policy.

Ultimately the downgrading of health in the SDGs is symptomatic of the time. However, we individually know what it is like to experience ill health ourselves or in our families. We know its importance and we are beginning to understand the concepts of well-being. The SDGs may be imperfect; fortunately, they can be adapted. They will not, however, allow us to consider the "spiritual" well-being of ourselves and our populations. This needs to be considered when talking of global health and target-setting.

Notes

1 *The New Testament*, Matthew 25:29, King James Version.
2 Letter from Angela Merkel, Nana Akugo-Addo and Erna Solberg to Dr Tedros, April 2018 www.bundesregierung.de/Content/DE/_Anlagen/2018/04/2018-04-19-brief-who-englisch.pdf?__blob=publicationFile&v=1. Accessed 21 September 2018.

References

Angenendt, S., and M. Voss. 2018. "Healthier Together: Global Health Governance in the Era of SDGs". *Council of Councils*, 28 June. Accessed 28 November 2018. www.cfr.org/councilofcouncils/global_memos/p39188.

Bloom, D. E., A. Khoury, and R. Subbaraman. 2018. "The Promise and Peril of Universal Health Care". *Science* 361 (6404), eaat9644.

Bradshaw, D., P. Wyk, N. Somdyala, and M. Cerf. 2016. *The Burden of Health and Disease in South Africa*. 15 March. Accessed 28 November 2018. http://pmg-assets.s3-website-eu-west-1.amazonaws.com/160315SAMRC_Burden.pdf.

Canadian Institute for Health Information. 2009. *Improving the Health of Canadians: Exploring Positive Mental Health*. Canadian Population Health Initiative. Ottawa. Accessed 28 November 2018. www.cihi.ca/en/improving_health_canadians_en.pdf.

Commission on the Social Determinants of Health. 2008. *Closing the Gap in a Generation: Health Equity through Action on the Social Determinants of Health: Final Report of the Commission in Social Determinants of Health*. Geneva: World Health Organization. Accessed 28 November 2018 http://apps.who.int/iris/bitstream/handle/10665/43943/9789241563703_eng.pdf;jsessionid=38514421C49957310F7C844C8039C71B?sequence=1.

Copenhagen Consensus Center. 2018. "Post-2015 Consensus: What Are the Smartest Targets for the Post-2015 Development Agenda?". Accessed 25 June 2018. www.copenhagenconsensus.com/post-2015-consensus.

Davies, S. C., E. Winpenny, S. Ball, T. Fowler, J. Rubin, and E. Nolte. 2014. "For Debate: A New Wave in Public Health Improvement". *The Lancet* 384 (9957): 1889–1895.

Frenk, J., and S. Moon. 2013. "Governance Challenges in Global Health". *The New England Journal of Medicine* 368 (10): 936–942.

Galatsidas, A., and F. Sheehy. 2015. "What Have the Millennium Development Goals Achieved?". *Guardian*, 6 July. Accessed 28 November 2018. www.theguardian.com/global-development/datablog/2015/jul/06/what-millennium-development-goals-achieved-mdgs.

Grépin, K. A. 2015. "Power and Priorities: The Growing Pains of Global Health: Comment on 'Knowledge, Moral Claims and the Exercise of Power in Global Health'". *International Journal of Health Policy and Management* 4 (5): 321–322.

IHME (The Institute for Health Metrics and Evaluation). 2018a. "Global Burden of Disease Comparisons". Accessed 18 June 2018. https://vizhub.healthdata.org/gbd-compare/.

IHME (The Institute for Health Metrics and Evaluation). 2018b. "Global Burden of Disease Study 2016 (DGB 2016): Socio-Demographic Index (SDI): 1970–2016". Accessed 20 September 2018. http://ghdx.healthdata.org/record/global-burden-disease-study-2016-gbd-2016-socio-demographic-index-sdi-1970-2016.

IHME (The Institute for Health Metrics and Evaluation). 2018c. "Millennium Development Goals (MDGs) Visualization: Are Countries on Track to Meet the Millennium Development Goals?" Accessed 22 June 2018. https://vizhub.healthdata.org/mdg/.

Kaasch, A. 2015. *Shaping Global Health Policy. Global Social Policy Actors and Ideas about Health Care Systems.* Basingstoke: Palgrave Macmillan.

Matheson, D. 2015. "Will Universal Health Coverage (UHC) Lead to the Freedom to Lead Flourishing and Healthy Lives? Comment on 'Inequities in the Freedom to Lead a Flourishing and Healthy Life: Issues for Healthy Public Policy'". *International Journal of Health Policy and Management* 4: (1): 49–51.

McArthur, J., and K. Rasmussen. 2017. "How Successful Were the Millennium Development Goals?". *Brookings*, 11 January. Accessed 19 June 2018. www.brookings.edu/blog/future-development/2017/01/11/how-successful-were-the-millennium-development-goals/.

Morrison, J. S. 2018. "The New Realism, An Uncomfortable Middle". *Center for Strategic and International Studies*, 17 August. Accessed 28 November 2018. www.csis.org/analysis/new-realism-uncomfortable-middle.

NCOP Social Services. 2016. "Burden of Health and Disease in South Africa: Medical Research Council Briefing". *Parliamentary Monitoring Group*, 15 March. Accessed 25 June 2018. https://pmg.org.za/committee-meeting/22198/.

Shiffman, J. 2014. "Knowledge, Moral Claims and the Exercise of Power in Global Health". *International Journal of Health Policy and Management* 3 (6): 297–299.

Stringhini, S., C. Carmeli, M. Jokela, M. Avendaño, P. Muennig, F. Guida, F. Ricceri, *et al.* 2017. "Socioeconomic Status and the 25×25 Risk Factors as Determinants of Premature Mortality: A Multicohort Study and Meta-analysis of 1.7 Million Men and Women". *The Lancet* 389 (10075), March 2015.

The Mental Health Foundation. 2017. "Surviving or Thriving? The State of the UK's Mental Health". Report #MHAW17 for Mental Health Awareness Week 2017.

London. Accessed 28 November 2018. www.mentalhealth.org.uk/publications/surviving-or-thriving-state-uks-mental-health.

UNGA (United Nations General Assembly). 2011. *2011 High Level Meeting on Prevention and Control of Non-communicable Diseases. General Assembly United Nations*, 19–20 *September 2011*. Accessed 25 June 2018. www.un.org/en/ga/ncdmeeting2011/.

UNGA (United Nations General Assembly). 2015. *Transforming our World: the 2030 Agenda for Sustainable Development.* A/RES/70/1. Accessed 28 November 2018. www.un.org/en/development/desa/population/migration/generalassembly/docs/global compact/A_RES_70_1_E.pdf.

United Nations Security Council. 2000. *Resolution 1308 (2000)* Adopted by the Security Council at its 4172nd meeting, on 17 July 2000. Accessed 21 June 2018. https:// undocs.org/S/RES/1308(2000).

Van de Pas, R., P. S. Hill, R. Hammonds, G. Ooms, L. Forman, A. Waris, C. E. Brolan, M. McKee, and D. Sridhar. 2017. "Global Health Governance in the Sustainable Development Goals: Is it Grounded in the Right to Health?" *Global Challenges (Hoboken, Nj)* 1 (1): 47–60.

Whiteside, A. 2014. "South Africa's Key Health Challenges". *The ANNALS of the American Academy of Political Science* 652 (1): 166–185.

WHO (World Health Organization). 2005. *Promoting Mental Health: Concepts, Emerging evidence, Practice.* A report of the World Health Organization, Department of Mental Health and Substance Abuse in collaboration with the Victorian Health Promotion Foundation and the University of Melbourne. Geneva: World Health Organization. Accessed 19 September 2018. www.who.int/mental_health/evidence/en/promoting_ mhh.pdf.

WHO (World Health Organization). 2006. *Constitution of the World Health Organization – Basic Documents.* 55th edn. Supplement, October 2006. Accessed 20 September 2018. www.who.int/governance/eb/who_constitution_en.pdf.

WHO (World Health Organization). 2018a. "Emergencies Preparedness, Response: Alert and Response Operations". *WHO.* Accessed 28 November 2018. www.who.int/csr/ alertresponse/en/.

WHO (World Health Organization). 2018b. "International Health Regulations (IHR)". *WHO.* Accessed 20 September 2018. www.who.int/topics/international_health_ regulations/en/.

WHO (World Health Organization). 2018c. "The Rio Political Declaration on Social Determinants of Health". Rio de Janeiro, Brazil, October 2011. Accessed 25 June 2018. www.who.int/sdhconference/declaration/Rio_political_declaration.pdf?ua=1.

WHO (World Health Organization). 2018d. "Third United Nations High-level Meeting on NCDs". *World Health Organization.* Accessed 4 September 2018. www.who.int/ ncds/governance/third-un-meeting/en/.

Wilkinson, R. G., and K. Pickett, K. 2010. *The Spirit Level: Why Equality is Better for Everyone.* London: Penguin.

Wilkinson, R. G., and K. Pickett. 2018. *The Inner Level: How More Equal Societies Reduce Stress, Restore Sanity and Improve Everyone's Well-being.* Allen Lane. Accessed 29 November 2018. www.penguinrandomhouse.com/books/567749/the-inner-level-by-richard-wilkinson-and-kate-pickett/9780525561224/.

World Bank. 1993. *World Development Report 1993: Investing in Health.* Oxford: Oxford University Press. Accessed 28 November 2018. https://openknowledge.worldbank.org/ handle/10986/5976.

World Bank. 2016. "World Ban Group Launches Groundbreaking Financing Facility to Protect Poorest Countries against Pandemics". *The World Bank*, 21 May. Accessed 28 November 2018. www.worldbank.org/en/news/press-release/2016/05/21/world-bank-group-launches-groundbreaking-financing-facility-to-protect-poorest-countries-against-pandemics.

4 Gender equality from the MDGs to the SDGs

The struggle continues

Sara Rose Taylor and Rianne Mahon

Introduction

Gender equality has been on the international agenda for almost a century. As early as the Treaty of Versailles, 1920, feminists succeeded in asserting a place for women in the League of Nations and the International Labour Organization (ILO) (Caglar, Prügl, and Zwingel 2013a). The United Nations Commission on the Status of Women (CSW), formed in the aftermath of World War II, subsequently served to push for women's equality in the UN. Perhaps one of its greatest achievements was the General Assembly's decision to declare 1975 International Women's Year, which served to launch the UN Decade for Women. The latter in turn contributed to the formation of a global feminist movement capable of influencing the outcome of a series of important UN conferences in the 1990s that served to establish gender equality and women's empowerment as global norms.

Forged in the wake of these conferences, the Millennium Development Goals (MDGs) reflected a rhetorical commitment to gender equality as one of the eight goals. Its measurement was however reduced to three indicators: gender disparities in education, women's share in non-agricultural wage employment, and the proportion of seats held by women in national parliaments.[1] In contrast, the Sustainable Development Goals (SDGs) offer a much richer definition of the gender equality goal and gender equality is mainstreamed through seven of the other 18 goals. The contrast between the two in part reflects the processes through which they were developed. Arguably, while the MDGs were decided by a small group of development technocrats in closed rooms (Fukuda-Parr and Hulme 2011), the formation of the SDGs provided a substantial space for contestation, which women's organizations and their allies utilized to make some important, if partial, gains. Nevertheless "the meaning of a norm derives not from an imagined, codified intention but from the process of norm translation", best seen as "fluid propositions that need to gain reception and pass through societal filters before they become part of the vernacular" (Caglar, Prügl, and Zwingel 2013b, 290). In other words, what becomes of the formal gains – and gaps – will be determined by subsequent struggles at local, national, regional, and global scales.

This chapter develops the argument as follows. The first section reviews debates among feminists about the advantages and disadvantages of participating in exercises like the SDGs that involve the quantification of global norms. We concur with those who argue that while quantification does pose certain risks, it would be a mistake to refuse to participate. What matters most is the utilization of contestation spaces wherever they exist. The second section compares the processes by which the MDGs and SDGs were arrived at, with a focus on the treatment of gender equality. While the former represented a retreat from the lively contestations surrounding the UN conferences in the first half of the 1990s, the latter opened up important spaces for contestation. The third section examines how the SDGs identify gender equality and assesses the gains and limitations this represents. Particular attention is paid to the issue of support for women's ongoing participation in defining what the SDGs mean in practice.

Governance by indicators: feminist debates

The inclusion of gender in the SDGs has been the subject of much discussion by feminists, particularly with regard to the use of indicators. Some concerns are methodological, some normative, and some a mix of the two. Merry's *The Seductions of Quantification* (2016), addresses many concerns voiced in feminist discussion of indicators, as do Liebowitz and Zwingel (2014). First, there is a disconnect between the seemingly objective nature of indicators. Recognition of the underlying subjectivity of indicators is the first step towards contesting, transforming, and reprogramming them. The process of selecting indicators is a critical juncture in the agenda formation process as it can either broaden or narrow a goal's interpretation and open or close the space for contestation (see Rose Taylor 2018). In the case of the SDGs, which retained the MDGs' structural blueprint, feminists could push back against the normative foundation of the original indicators and contest the scope of those newly selected.

Feminists have also criticized the way indicators remove context. While indicators are valued for their simplicity, at the same time they decontextualize and dehistoricize the information they convey. This suggests that indicators need to be supplemented by the introduction of qualitative knowledge to bring in local knowledges and circumstances. Feminists have also identified the importance of including a broader range of people in indicator creation. There is a role for feminist methodology here, including participatory decision-making. Powell (2016) argues that gender indicators are not reflective of a substantive approach to equality, in large part because an indicator condenses an array of information into one package, typically derived from easily quantifiable information. Gender indicators have not consistently addressed complex realities and they have tended to ignore the intersectionality of gender, class, race-ethnicity, and other forms of inequality among women (Esquivel 2016).

Despite these overarching concerns, many feminists see a place for indicators. Feminist economists have made important contributions to the development of gendered data, including the Gender Empowerment Measure. Caglar, Prügl, and Zwingel (2013b) also provide a compelling case for using indicators and other quantitative methodologies as a feminist strategy in international governance. Feminists working inside existing structures can use indicators as strategic frames for legitimizing gender equality. As Goetz and Jenkins argue "gender-specific targets 'call out' gender as a specific aspect of the positive change they propose, and thus note a gender dimension to the problem being addressed" (2016, 129). The indicator's role in framing understanding can thus be used to advantage to the extent that the data make visible the way gender equality policies have been implemented to better understand impacts and causal relations.

As Merry (2016) notes, as indicators become more established over time, it becomes more difficult to contest their underlying frameworks or categories of analysis. As we shall see, although feminists sought to challenge the MDGs' narrow definition of gender equality, they met with very limited success. The SDG formation process, however, provided an important opening to negotiate both the content of, and approach to, indicators. It offered an important contestation space; a space where indicators could be reformulated and the measurement approach could be restructured. The indicator-based structure remained intact, but as a result of the engagement of women's groups, the SDG indicators reflect a more human rights-based approach than the MDG indicators. Key feminist actors, including UN Women, the Post-2015 Women's Coalition and the Women's Major Group (WMG), played an important role in helping to reorient the underlying approach such that the SDGs came to embody many (though not all) of their demands.

From the MDGs to the SDGs: contrasting processes

The MDGs ostensibly emerged from the Millennium Development Summit, which committed member countries to a set of values and principles for the new century, including equal rights of all. Gender equality and women's empowerment were also acknowledged "as effective ways to combat poverty, hunger and disease and to stimulate development that is truly sustainable" (UN General Assembly 2000, 5). In fact, however, the eight MDGs emerged from a process that began earlier, in the wake a series of UN conferences which had provided important spaces for contestation for a vibrant transnational women's movement, uniting voices from North and South. Through these conferences, women activists succeeded in linking gender to environmental issues at the UN Conference on Environment and Development in Rio (1992), made women's rights integral to human rights at the UN Conference on Human Rights, Vienna (1993), and asserted the importance of sexual and reproductive rights at the UN Conference on Population and Development in Cairo (1994). This process culminated in the Fourth UN Conference for Women in Beijing (1995), where they won global endorsement for gender mainstreaming as well as "a comprehensive catalogue of state commitments in the Beijing Platform for Action" (Caglar, Prügl, and Zwingel 2013a).

Despite these gains, the transnational women's movement was not without its opponents. It was fiercely opposed by "the unholy alliance" of the Holy See, determined to block the right to abortion, and by Islamic allies concerned about adolescents' access to family planning (Kabeer 2015). While this alliance engaged in open opposition to the recognition of sexual and reproductive rights, some OECD countries worked covertly to limit the gains. An important forum for such opposition proved to be the *Groupe de Réflexion* established by the OECD's Development Assistance Committee (DAC) to develop guidelines for the twenty-first Century. Their first report, *Shaping the 21st Century: The Contribution of Development Cooperation* (1996), enumerated a set of development goals, including gender equality – reduced however to eliminating gender disparities in primary and secondary education by 2005 – and access to reproductive services for all through the primary health care system. While the DAC thus rejected the unholy alliance's position on reproductive health, the narrow definition of gender equality reflected the Japanese government's concern to water down the Beijing Platform of Action (Hulme 2009).

While the DAC successfully enlisted the support of the other key development agencies for its definition of global international development goals, Kofi Annan sought to regain control of the global agenda by launching a plan for the Millennium Assembly, appointing John Ruggie to author the report that would act as a centrepiece for the summit. However, missing from *We the Peoples: The Role of the United Nations in the 21st Century* (Annan 2000) were both gender equality and reproductive rights. Gender equality was brought back into the final Millennium Declaration, but only as a means to achieve the main goal (poverty reduction) and any mention of reproductive health rights was avoided as it remained unacceptable to some of the UN's membership (Hulme 2010).

However, neither *We the Peoples* nor the Millennium Declaration specified the goals and indicators that would become the MDGs. While UN agencies like the ILO and the United Nations Development Fund for Women (UNIFEM)[2] stayed out of the discussion, focusing instead on their own agendas (Fukuda-Parr, Yamin and Greenstein 2014), the key donor agencies seized the initiative. Their joint publication, *A Better World for All: Progress towards the International Development Goals* (IMF, OECD, UN and the World Bank Group 2000), reiterated the DAC's 1996 International Development Goals, including both gender equality in education and reproductive health. The final version – which omitted reproductive health – was produced by a smaller task force of experts from the DAC, the World Bank, the International Monetary Fund (IMF) and the United Nations Development Programme (UNDP) to formalize the goals (Fukuda-Parr and Hulme 2011). In addition to narrowing the definition of gender equality, the focus on the South ("development") also made it look as if gender inequality were a problem confined to the South. As the report of the expert group on structural policy constraints in achieving the MDGs to the 2013 Commission on the Status of Women noted:

The idea that the MDGs apply only to developing countries fails to con-sider development in a broad, accountable and universal way. There is an inconsistent approach to human rights within many countries of the Global North, promoting human rights as a foreign policy or development objective while taking repressive measures towards women's rights in their own borders.

(CSW 2013, 26)

The MDG process subsequently provided opportunities to contest the nar-rowing of the gender equality agenda. Thus, in the lead up to MDG + 5, one of the task forces set up under the direction of Jeffrey Sachs brought together an impressive array of representatives from the UN women's agencies, academe (Diane Elson and Gita Sen), international non-governmental organizations (INGOs) (e.g. Planned Parenthood), and civil society organizations (e.g. the People's Action Forum) as well as representatives from the World Bank, and the Inter-American Development Bank. Caren Grown and Geeta Rao Gupta from the International Center for Research on Women were two of the lead authors of *Taking Action: Achieving Gender Equality and Empowering Women* (UNDP 2005), which accepted the use of indicators to measure progress, but sought to provide better indicators for girls' education, women's employment, and women's participation. It also attempted to broaden the agenda by adding sexual and reproductive rights, infrastructure to reduce women's and girls' time burden, women's property and inheritance rights, and violence against women. At this juncture, however, women's advocates lacked sufficient support to do more than restore (Target 5.b) universal access to reproductive health.

Although the 2010 High-Level Plenary Meeting on the MDGs issued a reso-lution that referred back to the broader gender equality agenda laid out in the Beijing Platform of Action, it resulted in no further changes to the MDGs. It did, however, serve to launch the process of developing a post-2015 agenda. Ini-tially, the process operated along two tracks. First, the Secretary General set up a UN System Task Team co-chaired by the UN Department of Economic and Social Affairs (UNDESA) and UNDP staff, which produced *Realizing the Future We Want for All* (2012). Gender equality remained part of this agenda and the report acknowledged that this would involve reducing gender inequalities in the labour market and access to productive resources, protection of reproductive rights, and improved access to good quality health, education, and social protec-tion (UNDESA 2012).

The Secretary General also launched a broader consultative process in which civil society groups were invited to participate. In September 2012, UN Women and GEAR (Gender/Equality/Architecture/Reform)[3] met to strategize about how to ensure gender remained central to the post-2015 agenda, which resulted in the formation of the Post-2015 Women's Coalition. Working with Oxfam and other INGOs, the Coalition used the occasion of the 57th meeting of the Commission on the Status of Women (CSW) to lobby for maintaining gender equality as a standalone goal. Toward this end, the allies engaged in "policy dialogues with UN member states, informal and formal meetings with UN Women and government, prepared fact sheets to hand out to delegates and other NGO colleagues" (Rosche 2016, 115) and succeeded in getting the idea of a standalone goal into the final outcome document of this and the 58th meetings of the CSW. Their struggle also left an imprint on the High-Level Panel's report, *A New Global Partnership: Eradicate Poverty and Transform Economies through Sustainable Development* (2013), which acknowledged the need to maintain a standalone gender equality goal, while enlarging it to include prevention of all forms of violence against women, recognition of women's economic rights and the importance of universal sexual and reproductive health and rights.

While the Coalition was gearing up for the post-2015 agenda, the Women's Major Group, associated with the UN Sustainable Development Process since 1992, was fighting to defend sexual and reproductive rights at the Rio+20 meetings in 2012 (Gabizon 2016). The WMG was thus well-placed to seize the opportunity when the Secretary General decided to merge the post-2015 and sustainable development processes. This decision resulted in an even more open process (Fukuda-Parr 2019) as the Open Working Group – and its nine working groups, including a now much larger WMG – became a central locus of negotiations for Agenda 2030. The Open Working Group gave the WMG

> access to all meetings, including informal negotiations, not only as observers but also with allotted speaking slots in all sessions, as well as the right to comment on the negotiation document, to have our comments published on the UN websites, and present key recommendations with keynote speakers in the plenary sessions.
>
> (Gabizon 2016, 103)

Although only one hour a day was allotted to civil society participation and governments were not obliged to be present, the two groups successfully

> monitored the negotiations and communicated the ongoing discussion back to their membership. Using social media (mostly via Twitter) during the negotiations in New York was critical for highlighting the progressive, as well as regressive, language proposed by various Member States.
>
> (Tesfaye and Wyant 2016, 143)

When some countries argued that gender mainstreaming meant a standalone gender goal was no longer needed and others opposed the term "gender equality", seen as a means of promoting transgender and gay rights, the WMG and its allies were able to gain the support of a group of countries behind a statement supporting the standalone goal and mainstreaming (Gabizon 2016). The WMG also encountered ongoing opposition to embedding sexual and reproductive rights in the SDGs. In addition to "the unholy alliance" of the past, a majority of the African group opposed including "other status" (LGBT), supported the position that parents should have the right to choose which type of education to give their children, defined the term "family" as referring to a man and a woman, opposed abortion, and argued that states should implement goals in line with their cultural and religious values.

(Tesfaye and Wyant 2016, 138)

Again, the WMG was able to appeal to its allies and a group of 58 states, led by South Africa, issued a joint statement at the 13th Session of the Open Working Group (OWG) recognizing that "the respect, promotion and protection of sexual and reproductive health and rights for all" (High-Level Task Force for ICPD 2014).

Certainly, the UN Secretary General worked to open up a space for contestation in determining the SDGs, but it was the merger of the post-2015 and sustainable development processes that really opened up the space for the WMG to influence the outcome. The merger of the two processes also influenced the choice of the annual High Level Political Forums as the monitoring mechanism (Fukuda-Parr 2019). These meetings, and the associated side events, organized by the Women's Major Group and Feminist Alliance for Rights (successor to the Post-2015 Women's Coalition) provide a space for ongoing contestation. Of particular importance, the merger shifted the scope from the MDGs' focus on monitoring the South to encompass the whole of the globe. This helps to make visible the persistence of gender inequalities in the North as well as the South: across the globe women have less secure jobs, lower pay, carry a disproportionate share of unpaid care work, experience gender violence, and struggle to have their voices heard.

The SDGs: significant victories but the struggle continues

The WMG and its allies made good use of the more open contestation space that the SDGs provided once the centre of action shifted to the OWG. In addition to winning the battle to maintain gender equality as a standalone goal and deepening its definition, gender appears in one form or another in Goals 1, 2, 3, 4, 6, 8, 10, 13, 16 and 17. In this section we focus on several key issues: sexual and reproductive rights, gender-based violence,[4] paid employment, women's unpaid care work, and women's participation. Yet, as numerous feminist commentators have noted, these gains mean little if key "enabling conditions" – gender responsive, counter-cyclical macroeconomic policies, the formation of an international tax body to deal with various forms of tax-avoidance, the reform of international financial and trade institutions to create a stable environment, and regulation of the operations of transnational corporations – are not met. Thus, we also consider the extent to which the SDGs have attempted to grapple with these issues.

As we have seen, sexual and reproductive health and rights have been contentious issues since the Cairo Population and Development Conference, where the International Women's Health Coalition made important breakthroughs despite the opposition. Opposition from UN members had kept this out of the original MDGs, although feminists succeeded in getting reproductive health services – though not rights – added in 2005. Despite continued opposition, the SDGs ended up including universal access to sexual and reproductive health and rights (Target 5.6), and explicit reference to the 1994 Conference on Population and Development and the Beijing Platform for Action. Moreover, Target 3.7 states "By 2030 ensure universal access to sexual and reproductive health care services, including for family planning information and education, and the integration of reproductive health into national strategies and programmes" (UN General Assembly 2015, 16). However, the SDGs remain silent on legislation criminalizing abortion and there is no reference to education about LGBTQ rights.

While the MDGs were silent on gender-based violence, the SDGs include several references to preventing violence against women. Thus Target 5.2 would "eliminate all forms of violence against all women and girls in the public and private spheres, including trafficking and sexual and other types of exploitation" (UN General Assembly 2015, 18) and 5.3 seeks to eliminate all harmful practices including female genital mutilation. Target 11.7 refers to universal access to safe, inclusive and accessible, green and public spaces, in particular for women and children, and one of the associated indicators is the proportion of persons who are the victim of physical or sexual harassment, by sex, age, disability status, and place of occurrence in the previous 12 months. In addition, the indicators associated with Target 16.1 are: "proportion of population subjected to physical, psychological or sexual violence in the previous 12 months" and "proportion of population that feel safe walking alone around the area they live" (UN General Assembly 2017, 21). Thus, although protection of LGBTQ persons is not explicitly mentioned, these indicators potentially open a space for asserting their rights.

Whereas MDG 3 ignored women's contribution to rural economies, SDG Target 5.a is to "undertake reforms to give women equal rights to economic resources, as well as access to ownership and control over land and other forms of property, financial services, inheritance and natural resources, albeit 'in accordance with national laws'" (UN General Assembly 2015, 18). The chosen indicator (5.a.1) calls for the identification of the proportion of total agricultural population with ownership or secure rights over agricultural land by sex and share of women among owners or rights-bearers of agricultural land by type of tenure. However, the SDGs still fail to adequately recognize that

> small farmers, particularly women farmers, pastoralists, artisanal fishermen and women and other small food providers are already feeding the majority of the world population and are more productive per unit [of land] than large industrial agriculture, while maintaining the largest seed and livestock diversity.
>
> (WMG 2014, 3–4)[5]

While the MDGs reduced women's paid employment to their share of non-agricultural employment, SDG 8 deals with gender inequalities in paid employment, notably the high share of women in informal employment and the persistence of wage inequalities even within the formal sector. Thus indicator 8.3.1 calls for the identification of the proportion of informal employment in non-agricultural employment by sex. Target 8.5 looks to achieve "full and productive employment and decent work for all women and men, including young people and persons with disabilities, and equal pay for work of equal value" (UN General Assembly 2015, 19) while Target 8.8 would "protect labour rights and promote safe and secure working environments for all workers, including migrant workers, in particular women migrants, and those in precarious employment" (UN General Assembly 2015, 20). Target 1.3 is also relevant as it aims at implementing "nationally appropriate social protection systems and measures for all, including floors, and by 2030 achieve substantial coverage of the poor and the vulnerable" (UN General Assembly 2015, 15). Together these targets and indicators represent a potentially significant advance over the MDGs especially in combination with Target 10.3 which would "ensure equal opportunity and *reduce inequalities of outcome* including by eliminating discriminatory laws, policies and practices and promoting appropriate legislation, policies and action in this regard" (UN General Assembly 2015, 21, emphasis added). However, as Rai, Brown and Ruwanpura caution,

> SDG 8 cannot … be successful in delivering "decent work for all" unless economic growth is decentred, paid employment as well as unpaid social reproductive work are recognized as important contributions to society and properly recompensed through state supported mechanisms, non-state initiatives and through cooperative and community actions.
>
> (2019, 371)

Although proponents failed to insert a standalone goal on migration (Crush, Chapter 6), migrants are mentioned in two of the goals and seven of the targets including 5.2, 8.7, 8.8 and 10.6. The SDGs

> have the potential to address the key factors that underline migration from a rights perspective if their realization goes beyond the provision of "safe and orderly migration" pathways to also address the lack of "decent work" and participation in decision-making from a (gendered) workers' rights perspective at all stages of migration.
>
> (Piper 2017, 231)

However, the tension between the "managed migration" approach and a rights-based approach is reflected in the SDGs. In Chapter 5 of this volume, KC and Hennebry discuss the gendered implications of this tension.

Goal 5 calls for the "recognition and valuation" of unpaid domestic work, ignored in the MDGs although feminists had been calling for its recognition for decades. Target 5.4 calls for "the provision of public services, infrastructure and social protection policies and the promotion of shared responsibility within the household and the family" – albeit, only "as nationally appropriate". The only indicator associated with this target, however, is the proportion of time spent on unpaid domestic and care work by sex, age and location. While the adoption of time-use surveys has shown this can be helpful, the indicator says nothing about the services required to redistribute care work. The achievement of other goals like Goal 6 on clean water and sanitation could go a long way toward lightening the unpaid labour of women in rural areas of the South. Target 4.2 dealing with early child development might have been helpful, but as currently worded it remains preoccupied with school-readiness and ignores the potential contribution of early childhood education and care programs to redistributing women's care work. Another indicator that is needed to promote the redistribution of women's unpaid care work is attitudes around men's equal role in care responsibilities (Rosche 2016).

Health care is another sector in which public investment can help redistribute women's unpaid care burden. Nevertheless, while Target 3.c calls for a substantial increase in health financing and recruitment, development, training, and retention, it limits its focus to developing countries, ignoring the way cuts to the health care systems of the North add to unpaid care work there too. Given the inadequacy of investment in health and other care services, migrant women from poorer countries have stepped in to fill care gaps. Here SDG 8.8, which explicitly refers to the labour rights of women migrants and the associated indicator (8.8.2), which refers to compliance of national labour legislation to ILO "textual sources" could help to bolster the struggle to get more countries, especially receiving countries, to ratify these conventions. Finally, the SDGs are virtually silent on the need to enshrine the right for "time to care" in the form of paid parental and other care-related leave.

Realizing any of the above gains depends on the existence of favourable "enabling conditions". One of these is a gender-responsive macroeconomic policy. Current macroeconomic policies typically do not create these conditions, as they are incorrectly seen as gender-neutral,[6] leading them to neglect the different costs men and women bear in the face of economic shocks. This includes opportunities for paid employment, demands for the recognition of women's unpaid labour, and changing resources for policies to reduce inequalities (Heintz 2015). In fact,

> countercyclical macroeconomic policies that are gender sensitive can expand social policy expenditure in critical areas for women, to prevent women's employment from becoming even more precarious, and to mitigate the impacts of food price speculation for small producers (many of whom are women) and poor households.
>
> (Bidegain Ponte and Rodríguez Enríquez 2016, 88)

Such policies require stable financial institutions at the global scale, global initiatives to inhibit tax avoidance, the regulation of the operations of transnational corporations, reforms to the global trade regime, and the introduction of new financing mechanisms including some form of a global tax.

The WMG managed to get some language into the SDGs on this. SDG 10.4 refers to the adoption of policies, "especially fiscal, wage and social protection programs"; however, there is only one indicator associated with it – labour's share of GDP – which does not adequately measure policy commitments or effectiveness. Target 17.13 calls for enhanced global macroeconomic stability through policy co-ordination,[7] while Target 10.5 encourages improvement of the regulation of global financial markets. However, a key associated indicator – "adoption of global financial transaction tax" i.e. a Tobin tax – was dropped in favour of "financial soundness indicators" developed by the IMF (Global Policy Watch 2016, 3). The SDGs hold no promise of regulating the operations of transnational corporations. They actively encourage public–private partnerships yet have not "established an appropriate monitoring body to ensure that these partnerships adhere to human rights standards and are in line with age, disability and gender equality and non-discrimination" (Tesfaye and Wyant 2016, 142).[8] More broadly, the SDGs and the associated Addis Ababa Action Agenda are largely silent on three core issues:

> 1) the process at the UN Human Rights Council to develop an international, legally-binding instrument to regulate, in international human rights law, the activities of transnational corporations and other businesses; 2) the need to establish mandatory, ex ante and periodic human rights and gender equality assessments of all trade and investment agreements; and 3) the importance of reviewing investor-state dispute settlement clauses to ensure that the right of states to regulate in critical areas for sustainable development is protected.
>
> (Bidegain Ponte and Rodríguez Enríquez 2016, 89)

Secure sources of funding constitute another enabling condition, which is addressed in SDG 7 as part of financing requirements and multi-stakeholder partnerships. Mobilizing funding from states as well as civil society partnerships will help address the issues identified above, enabling action instead of keeping the goals simply aspirational. Funding opportunities from within the UN system include the UN Trust Fund on Violence Against Women. It will however be important to tap into additional resources from outside the UN. UN Women's 2018 report on the SDGs' progress on gender equality provides examples for state-led financial investments in women, including gender-responsive budgeting, and funding led by donors and international organizations.

We have left one key aspect of the gender equality goal – women's participation – to the last precisely because it acts as the key "enabling" condition. While women's participation appeared as an indicator in the MDGs, it was limited to women's representation in national parliaments. The indicators currently chosen for Goal 5 simply add women's representation in local governments and managerial positions. UN Women however had proposed a broader target focused on strengthening women's collective action (Goetz and Jenkins 2016). This is important for although

> there is no homogeneous unified "women's movement" and that women's organizations do not all have the same agenda, they also agreed that strong women's movements, and their participation in and voice in policy-making and agenda-setting processes on the post 2015 agenda are imperative for the achievement of gender equality in the future agenda.
>
> (CSW 2013, 24)

The WMG and post-2015 Coalition also invoked Security Council Resolution 1324 to push for "targets on improving women's engagement in conflict resolution and prevention, demilitarising societies, and mitigating the impact of conflict and state fragility on women and girls", but "none of these survived the merger of the peace and governance goals" (Goetz and Jenkins 2016, 133).

Women's participation remains critical to the implementation of the SDGs. How wide a contestation space women's groups will have will partly be determined at the national (and local) scales: to what extent is the process for developing a particular country's voluntary report to the High Level Political Forum (HLPF) inclusive of women's voices and not just those of elite women's groups? According to a WMG survey of women's groups and networks, thus far the degree of inclusivity has been highly uneven. One-third of the groups had not been involved in any way, while slightly more than one-third had access to the official report and were engaged in drafting some parts of the report or asked to verify some information (WEDO 2017).

It also will depend on how regional reporting spaces are structured and the extent to which they represent open contestation spaces. Of particular importance are the annual meetings of the HLPF, which at present offer a contestation space that is somewhat constrained. The WMF and the Feminist Alliance for Rights (FAR) have been active at the forums held thus far. Both organizations however were disappointed with the opportunities currently provided for contestation. As WEDO, one of WMG's key members, noted:

> We expect the HLPR to be institutional and bureaucratic, but we also expect that the United Nations – as the only multilateral space and one based on human rights and peace – to uphold and fight for civil society space to bring ideas, support, critiques and accountability in all forms. WEDO firmly stands with the women's rights advocates and civil society colleagues who will be demanding changes in the coming year of HLPR to better serve the expansive 2030 Agenda to leave no one behind and to respect human rights and ensure justice for everyone – every step of the way.
>
> (WEDO 2017)

Along with other groups, WMF and FAR are pushing for a more open process that allows more time for discussion of the voluntary reports, thematic reviews and SDG implementation reviews.

One further area that needs to be opened to contestation is the determination of indicators. While some were relatively uncontentious, some areas have yet to be settled and others can be challenged. At present the task of developing indicators has been left to the UN Statistic Commission and the Inter-Agency Expert Group; however, much of the discussion is taking place behind closed doors among statisticians rather than experts in the thematic areas (Fukuda-Parr 2019, 17). It is important for women's groups to challenge their work. To do so however they need the "technical capacity to gender analysis of macroeconomic policy, as well as for engaging in indicator-based monitoring and evaluation processes … to uncover the political nature of these seemingly technocratic processes, and challenge them" (Esquivel 2016, 19).

Conclusions

Despite the shortfalls, the SDGs included some significant victories for the feminist agenda. Women's groups and their allies were able to use the contestation space provided by the Open Working Group to address many of the lacunae of the MDGs. Thus, the SDGs not only maintained a standalone goal on gender equality but also mainstreamed gender concerns throughout many of the other goals. Numerous important issues outlined above, notably absent from the MDGs, claimed space on the list of SDG targets and indicators. Disaggregating indicator data by a variety of demographic characteristics will also help to make visible the way gender intersects with other social relations to deepen inequalities.

Continued participation by women's groups is essential to preserve the contestation space and assert their role in interpreting the Goals. Women's groups can use many of the targets in SDG 5 and other goals to push for change in their own countries, in regional meetings and through the annual meetings of the HLPF, especially if WMG and its allies succeed in making those meetings more open to dialogue with civil society groups. In these battles, they will need both to defend their gains and to push beyond them. For sexual and reproductive health and rights, it is important to fight to include decriminalization of abortion and the rights of LGBTQ persons. This includes strengthening measures to prevent violence against women and extending these to cover all gender-based violence. It could also include engaging with the Yogyakarta Principles (ICJ 2007, 2017) for applying international human rights law in relation to sexual orientation and gender identity, strategically engaging with Targets 11.7 and 16.1 and integrating a process which runs parallel to the SDG process and uses some similar language.

The SDGs offer potential support for struggles to improve the economic position of rural and urban women, though the recent wave of disclosures suggests that at a minimum one indicator – incidence of sexual harassment in the workforce – needs to be added. Moreover, too often the concerns of women with disabilities are overlooked (Shakespeare and Williams forthcoming). The WMG and FAR should push for strengthening women's collective action perhaps using

Oxfam's proposed indicator, "proportion of women in autonomous women's organizations, networks and movements" (Rosche 2016, 118). There is also potential for strategically framing avenues of progress towards the creation of other important enabling conditions. For example, gender-responsive macro-economic policy can help to cement alliances with other actors. These examples all point to the importance of alliances with a variety of women's organizations that represent experiences across intersecting inequalities. Gender cuts across many areas in the SDGs; so too must advocacy and alliances with diverse groups.

Notes

1 Goal 5 also included improve maternal health. In 2005 this was broadened to include universal access to reproductive health.
2 UNIFEM is now part of UN Women.
3 GEAR was organized in 2008 as a feminist alliance to push for a strong and unified representation of women in the UN system.
4 We use this term rather than violence against women as it can also include violence perpetrated in the name of sexual identity.
5 For more on this see Blay-Palmer and Young, Chapter 2.
6 For a comprehensive overview of gender-blindness in macroeconomic policy, see Elson (2002).
7 The associated indicator is a "macroeconomic dashboard". At the time of writing, the indicators included in the dashboard have not been released, but in general it will aim to capture the complexities of the global economy. Measuring global macroeconomic stability would allow countries to pursue policies to protect themselves from potential crises in times of overall instability. The dashboard approach will yield more information than using a single indicator and would address Target 17.19 (build measurements that complement Gross Domestic Product).
8 See Horton, Chapter 13, for a more positive view of such partnerships.

References

Annan, K. A. 2000. *We the Peoples: The Role of the United Nations in the 21st Century*. Report of the Secretary General to the 54th Session. Agenda item 48 (b) The Millennium Assembly of the United Nations.

Bidegain Ponte, N., and C. Rodríguez Enríquez. 2016. "Agenda 2030: A Bold Enough Framework Towards Sustainable, Gender-just Development?" *Gender and Development* 24 (1): 83–98.

Caglar, G., E. Prügl, and S. Zwingel. 2013a. "Introducing Feminist Strategies in International Governance". In *Feminist Strategies in International Governance*, edited by G. Caglar, E. Prügl, and S. Zwingel, 1–18. Abingdon: Routledge.

Caglar, G., E. Prügl, and S. Zwingel. 2013b. "Conclusion: Advancing Feminist Strategies in International Governance". In *Feminist Strategies in International Governance*, edited by G. Caglar, E. Prügl, and S. Zwingel, 283–294. Abingdon: Routledge.

CSW (Commission on the Status of Women). 2013. *Strategic Policy Constrains in Achieving the MDGs on Women and Girls*. Expert Group Report prepared by R. Balakrishnan and V. Esquivel for the 58th meeting of the Commission on the Status of Women. EGM/MDG/2013/Report.

DAC (Development Assistance Committee). 1996. *Shaping the 21st Century: The Contribution of Development Cooperation.* Paris: OECD. Accessed 13 October 2018. www.oecd.org/dac/2508761.pdf.

Elson, D. 2002. "Macroeconomics and Macroeconomic Policy from a Gender Perspective". *Public Hearing of Study Commission on Globalization of the World Economy-Challenges and Responses.* Berlin: Deutscher Bundestag.

Esquivel, V. 2016. "Power and the Sustainable Development Goals: A Feminist Analysis". *Gender and Development* 24 (1): 9–23.

Fukuda-Parr, S. 2019. "Sustainable Development Goals (SDGs) and Global Goals". In *Oxford Handbook on the United Nations, Oxford Handbooks Online,* edited by S. Daws and T. G. Weiss. 2nd edn. Oxford: Oxford University Press.

Fukuda-Parr, S., and D. Hulme. 2011. "International Norm Dynamics and 'The End of Poverty': Understanding the Millennium Development Goals". *Global Governance* 17 (1): 17–36.

Fukuda-Parr, S., A. E. Yamin, and J. Greenstein. 2014. "The Power of Numbers: A Critical Review of MDG Targets on Human Development and Human Rights". *Journal of Human Development and Capabilities* 15 (2–3): 105–117.

Gabizon, S. 2016. "Women's Movements' Engagement in the SDGs: Lessons Learned from the Women's Major Group". *Gender and Development* 24 (1): 99–110.

Global Policy Watch 2016. *2030 Agenda and the SDGs: Indicator Framework, Monitoring and Reporting.* Accessed 13 October 2018. www.globalpolicywatch.org/blob/2016/03.18/2030-agenda-sdgs-indicator/.

Goetz, A. M., and R. Jenkins. 2016. "Gender, Security and Governance: The Case of SDG 16". *Gender and Development* 24 (1): 127–137.

Heintz, J. 2015. *Why Macroeconomic Policy Matters for Gender Equality.* UN Women, Policy Brief No. 4. Accessed 13 October 2018. www.unwomen.org/-/media/headquarters/attachments/sections/library/publications/2015/unwomen-policybrief04-macroeconomicpolicymattersforgenderequality-en.pdf?la=en&vs=349.

High-Level Panel of Eminent Persons on the Post-2015 Development Agenda. 2013. *A New Global Partnership: Eradicate Poverty and Transform Economies through Sustainable Development.* New York: United Nations.

High-Level Task Force for ICPD. 2014. "Task Force Welcomes the Joint Statement Made by South Africa on Behalf of 58 Member States at the Sustainable Development Goals Open Working Group". *High-Level Task Force for ICPD,* 14–18 July. Accessed 16 October 2018. http://icpdtaskforce.org/the-task-force-welcomes-the-joint-statement-by-58-member-states-at-the-sustainable-development-goals-open-working-group/.

Hulme, D. 2009. *The Millennium Development Goals (MDGs): A Short History of the World's Biggest Promise.* Brooks World Poverty Institute. Working Paper 100. Accessed 13 October 2018. www.unidev.info/Portals/0/pdf/bwpi-wp-10009.pdf.

Hulme, D. 2010. "Lessons from the Making of the MDGS: Human Development Meets Results-Based Management in an Unfair World". *IDS Bulletin* 41 (1): 15–25.

ICJ. 2007. *Yogyakarta Principles – Principles on the Application of International Human Rights Law in Relation to Sexual Orientation and Gender Identity.* Yogyakarta: International Commission of Jurists.

ICJ. 2017. *The Yogyakarta Principles plus 10.* Geneva: International Commission of Jurists. Accessed 13 October 2018. https://yogyakartaprinciples.org/wp-content/uploads/2017/11/A5_yogyakartaWEB-2.pdf.

IMF, OECD, UN and the World Bank Group. 2000. *A Better World for All: Progress Towards the International Development Goals*. Washington, DC: IMF, OECD, UN, and World Bank.

Kabeer, N. 2015. "Tracking the Gender Politics of the MDGs: Struggles for Interpretive Power in the International Development Agenda". *Third World Quarterly* 36 (2): 377–395.

Liebowitz, D. J., and S. Zwingel. 2014. "Assessing Global Gender (In)Equality: A CEDAW-Based Approach". Paper presented at the *International Studies Association Annual Convention*, Toronto, 26–29 March.

Merry, S. E. 2016. *The Seductions of Quantification*. Chicago: The University of Chicago Press.

Piper, N. 2017. "Migration and the SDGs". *Global Social Policy* 17 (2): 231–238.

Powell, C. 2016. "Gender Indicators as Global Governance: Not your Father's World Bank". *Georgetown Journal of Gender and the Law* 17 (3): 777–807.

Rai, S, B. Brown, and K. Ruwanpura. 2019. "SDG8: Decent Work and Economic Growth: A Gendered Analysis". *World Development* 113: 368–380.

Rosche, D. 2016. "Agenda 2020 and SDGs: Gender Equality at Last? An Oxfam Perspective". *Gender and Development* 24 (1): 111–126.

Rose Taylor, S. 2018. "On Measurement and Meaning: How Indicators Shape, Make, or Break Global Policy Goals and Outcomes". *The Role of Indicators in Promoting Gender Equality Through the Millennium and Sustainable Development Goals*, PhD diss., Wilfrid Laurier University.

Shakespeare, T., and F. Williams. Forthcoming. *Care and Assistance: Issues for Persons with Disabilities, Women and Care Workers*. International Labour Organization Report.

Tesfaye, S., and R. Wyant. 2016. "Achieving Gender Equality and Empowering All Women and Girls". In *International Norms, Normative Change and the UN Sustainable Development Goals*, edited by N. Shawki, 131–150. Lanham: Lexington Books.

UNDESA. 2012. *Realizing the Future We Want for All*. Report to the Secretary General of the UN Task Team on Post-2015 Development Agenda. New York: United Nations.

UNDP. 2005. *Taking Action: Achieving Gender Equality and Empowering Women*. London: United Nations Development Programme. Accessed 13 October 2018. www.undp.org/content/dam/aplaws/publication/en/publications/poverty-reduction/poverty-website/taking-action-achieving-gender-equality-and-empowering-women/Taking%20Action-%20Achieving%20Gender%20Equality%20and%20Empowering%20Women.pdf.

UN General Assembly. 2000. Resolution 55/2, *United Nations Millennium Declaration*, A/RES/55/2, 8 September. Accessed 16 October 2018. www.un.org/millennium/declaration/ares552e.htm.

UN General Assembly. 2015. Resolution 70/1, *Transforming our World: The 2030 Agenda for Sustainable Development*, A/RES/70/1, 21 October. Accessed 16 October 2018. https://undocs.org/A/RES/70/1.

UN General Assembly. 2017. Resolution 71/313, *Work of the Statistical Commission Pertaining to the 2030 Agenda for Sustainable Development*, A/RES/71/313, 10 July. Accessed 16 October 2018. https://undocs.org/A/RES/71/313.

UN Women. 2018. *Turning Promises into Action: Gender Equality in the 2030 Agenda for Sustainable Development*. New York: United Nations.

WEDO. 2017. "Finding Feminist Inspiration at an Uninspiring HLPF". *WEDO*, 4 August. Accessed 16 October 2018. www.wedo.org/finding-feminist-inspiration-uninspiring-hlpf.

WMG (Women's Major Group). 2014. *Women's "8 Red Flags" Following the Conclusion of the Open Working Group on Sustainable Development Goals.* Accessed 13 October 2018. www.womenmajorgroup.org/wp-content/uploads/2014/07/Womens-Major-Group_OWG_FINALSTATEMENT_21July.pdf.

WMG (Women's Major Group). 2017. *High Level Political Forum – Position Paper Executive Summary.* Accessed 13 October 2018. https://sustainabledevelopment.un.org/content/documents/14774HLPF_WMG_Paper_2017_final_UPDATED.pdf.

5 Gender, labour migration governance, and the SDGs

Lessons from the case of Nepal

Hari KC and Jenna L. Hennebry

Introduction

Women constitute above 48 per cent of the global migrant population (UN 2017), with growing numbers migrating explicitly for the purposes of work. Women, particularly those from the less developed countries facing deteriorating economic conditions, migrate to the developed countries which face an increasing "care deficit" (Peterson 2012). The "feminization" of work and migration has thus shifted the international division of reproductive labour from middle-class women of the developed countries to women migrants from the "peripheral" countries (Kilkey, Perrons, and Plomien 2013). However, women migrants' "feminized" skills in the care industry are not only poorly paid but also devalued and demeaned (Oksala 2013). In the global economy, women migrants are therefore confronting precarious situations on many fronts. Precarity multiplies when the "feminized" domestic work is performed by temporary women migrant workers. Lan (2008, 835) argues that domestic labour migration is akin to "colonial encounters" and has formed "interior frontiers" "built within the national frontier and in the intimate spheres of marriage and domesticity". Succinctly put, women migrants engaged in global labour markets are thus increasingly subjected to the dictates of neoliberal governmentality that exploits women through the feminization, flexibilization, and informalization of work (Peterson 2012).

In this chapter, we use the case study of Nepal to look at the interactions of the Sustainable Development Goals (SDGs) as they apply to women migrant workers. We explore three dimensions of the issue in question. First, we discuss how, despite the fact that there is a lack of a single SDG devoted to migration, the inclusion of migration and particularly that of women migrants in the SDGs fills a void in migration governance at the global level and thus bears important significance particularly in the context of poor countries such as Nepal, where migration provides a survival strategy for many women in the face of various intersecting local and non-local oppressive systems. Second, we argue that the SDGs conceive women's migration through a depoliticized lens, even though migration is a highly gendered and political process. This raises a question as to whether the SDGs address the entrenched socio-economic-political structures of gender inequality embedded in the other governance paradigms, including those of women's migration in both the migrant-sending and migrant-receiving countries, and whether they provide the necessary tools to address systemic discriminations and exploitative practices at their root. Finally, since women's labour migration is deeply intertwined with sustainable human development, we provide some suggestions on ways to address the challenges of implementing the SDGs so as to achieve substantive gender equality and women migrant workers' empowerment as envisaged in the SDGs.

SDGs and women's labour migration

Though not legally binding and thus largely dependent on the political will and moral commitments of states, the SDGs redress the silence on migration in the preceding Millennium Development Goals (MDGs) by recognizing "the positive contribution of migrants for inclusive growth and sustainable development" and by noting that "international migration is a multidimensional reality of major relevance for the development of countries of origin" (UNGA 2015, 8). Taylor and Mahon (Chapter 4) note that while the MDGs just provided a rhetorical commitment to gender equality, the SDGs offer a much richer definition that is mainstreamed through seven of the other 18 goals. Even the process through which the SDGs were formulated, by extensively engaging with civil society representing the lived realities of the people and particularly with the middle-income member states, was laudable (Esquivel and Sweetman 2016). The SDGs recognize that women's oppression is grounded in structural forces and institutions, both public and private, and that gender discrimination is deeply embedded in power inequalities and discriminatory norms that cut across social, economic, and political areas (Razavi 2016). Further, the SDGs have broadened the scope of the targets under the goal on gender equality and women's empowerment, and recognized that gender equality has social, economic, and political dimensions. As a comprehensive standalone goal on gender equality, SDG 5 seeks to achieve gender equality and empower all women and girls.

Although migration is specifically mentioned in five goals (5, 8, 10, 16, and 17), the SDGs fail to make room for a single goal focused on migration, despite much prior discussion on it (see Crush, Chapter 5). Crush contends that the drafters of the SDGs relegated migration to a few general targets scattered among the SDGs, instead of providing a single SDG on international migration. The dispersal of migration all across the SDGs, Crush states, explains why migration remained entirely excluded from the MDGs. What is to be noted, however, is that the 2030 Agenda clearly recognizes people's rights to movement and the need for the prevention of trafficking and forced labour, the protection of labour, and the facilitation of safe mobility. The interaction between gender and migration is relevant in a number of other goals and indicators. SDG 5a states the need to "undertake reforms to give women equal rights to economic resources, as well as access to ownership and control over land and other forms of property, financial services, inheritance and natural resources", albeit "in accordance with national laws" (UNGA 2015, 18). Target 5.2 in particular asks concerned parties to "[e]liminate all forms of violence against all women and girls in the public and private sphere, including trafficking and sexual and other types of exploitation" (UNGA 2015, 18). Targets 10.2 and 10.4 focus on promoting social, economic, and political inclusion for all, and the adoption of policies that progressively achieve gender equality. Specifically, Target 5.4 emphasizes the need to "[r]ecognize and value unpaid care and domestic work through the provision of public services, infrastructure and social protection policies and the promotion of shared responsibility" (UNGA 2015, 18). With respect to women migrants this would involve: recognizing the value of global care chains in countries of origin and destination; identifying the role and costs of transnational mothering; challenging and social constructions that undermine the value of women migrants and their contributions through paid and unpaid work.

Further, Goal 8 on "Decent Work and Economic Growth" is relevant with respect to women and migration, particularly Target 8.5 (ensuring decent work and equal pay for equal work), Target 8.7 ("eradicate forced labour, end modern slavery and human trafficking"), and Target 8.8, which specifically addresses the need to "protect labour rights and promote safe and secure working environments for all workers, in particular women migrants, and those in precarious employment"(UNGA 2015, 20). Target 10.7 focuses on facilitating "orderly, safe, regular and responsible migration through the implementation of planned and well-managed migration policies" (UNGA 2015, 21).

Esquivel and Sweetman (2016) argue that the principles outlined in the strongly aspirational Agenda 2030 are worlds away from the lived realities of the poorest and most marginalized individuals living in the Global South, and that they may not be translatable into actions that would unequivocally benefit the people they are targeted at helping. The Agenda has an over-ambitious vision which is not supported by strong enough language, clear means of implementation and policies, or funding mechanisms (Esquivel and Sweetman 2016). Another major obstacle that may affect the success of the Agenda 2030 is its implementation, which will be the responsibility of local governments and civil society actors. Given that the SDGs and targets are global in nature, individual states will ultimately be responsible for their implementation, provided that said states' policy spaces are protected (Esquivel 2016). The gender inequalities in income, wealth, and power have been produced and reproduced at national and global levels through the actions of powerful actors, especially big transnational corporations that have caused the very problems the SDGs are trying to solve (Esquivel 2016). Similarly, Fukuda-Parr (2016) fears that the transformational potential of the Agenda will be lost in the implementation process through selectivity, simplification, and national adaptation. Since the SDGs ask states to act "in accordance with national laws" (UNGA 2015, 18), countries will neglect those goals and targets that address the need to challenge power relations, reform institutions, and achieve other changes in the structures of political economic and social life, given that these will be the hardest to implement and achieve. This explicates the complexities of tackling the structural and everyday inequalities experienced by women migrant workers.

The principal onus of localizing the SDGs relating to gender equality and particularly to women migrant workers lies with the states as the main actors in the governance of migration. In the context of women migrants migrating to the more developed countries for work, three challenges starkly stand out. First, there is no gender-sensitive migration governance model, which results in women's disempowerment. Second, since the issue of women migrant workers is not an isolated subject, but it intersects with many other social-economic-political systems of domination, discriminatory policies, and institutional practices, addressing those interlocking systemic nexuses is a daunting task. Third, in the context of low- and middle-income countries, simply formulating gender-responsive policies would not suffice to effect a positive change in the lived experiences of women migrants due to the serious governance challenges of implementation.

Women migrants and gendered labour migration governance: the case of Nepal

A large segment of Nepal's migrant population comprises women who originate mainly from Nepal's newly urbanizing places and remote rural villages, where most people have adopted mixed livelihood strategies that combine subsistence farming, livestock, and the extraction of local natural resources. While most Nepali women have been traditionally engaged in unpaid care work and seasonal informal agricultural sectors, the penetration of unregulated global markets even in the smallest of the villages is exerting tremendous pressure on women's traditional livelihoods, lifestyles, and aspirations that now demand a monetary mode of production. There is thus a disjuncture between the mode of production Nepali women have been involved in and the capitalist monetary mode of production required to address the rapidly transforming realities wrought by globalization. This disconnect operates as a powerful catalyst pushing women out of the villages. Nepali women's migration for employment abroad thus results from an interplay between various intersecting and interlocking pre-existing systems of oppressions, and the newly emerging pressures from global markets and capitalism.

Most women migrating to the Persian Gulf countries for employment come from extremely impoverished situations such as those faced by Nepali women. These women have the choice of the "un-freedom of poverty" or the "un-freedom of servitude", and for them, the latter is a much better deal (Sunam 2014). The unemployment of men adds further pressure on women to find ways to ensure household survival, which, coupled with the dominant patriarchy, produces heightened risks of physical and psychological violence against women. For many Nepali women, labour migration has become a means to escape gender violence (Massey, Axinn, and Ghimire 2010; ILO 2015a; Kharel 2016).

The three major policies regulating Nepal's labour migration include the Foreign Employment Policy (2011), the Foreign Employment Act (2007), and the Foreign Employment Regulation (2008). These policies do not, in theory, discriminate between men and women seeking employment abroad (Sijapati 2014). However, in practice, Nepal's migration policy treats men's and women's mobility very differently. Nepal's labour migration policy is not based on a broader framework for maximizing the economic impacts of migration while also securing the protection of migrant workers (ILO 2015a). Existing policies as well as bilateral agreements Nepal has signed with some destination countries are focused more on regulating the labour migration process than on harnessing the benefits of migration and protecting migrant workers (ILO 2015a).

Since most Nepali women migrants are engaged in domestic work in the Middle East, which remains highly unregulated, various forms of abuses and exploitation remain largely ignored. Women face systematic disadvantages, such as employment in informal jobs, and the work in the unpaid care economy (UNGA 2015). Besides extreme forms of physical violence reported by the media, many acts of violence against Nepali women migrants in the Gulf countries are perpetrated by individual employers in non-physical forms such as low wages, withholding payments, or non-payment, confiscation of passports, overwork, constant nagging, inadequate rest, and the overall feeling of confinement (Kharel 2016). The Gulf Cooperation Council (GCC) countries in particular have granted their citizens unbridled power over migrant domestic workers with impunity that Johnson and Wilcke call (2015, 135) a "structural violence"; it is a "state-produced and -sanctioned relation" between migrants and employers that consolidates the citizens' power and control over the women migrant domestic workers.

When media reports on abuses and exploitation of Nepali women migrant workers abroad trigger an uproar and resistance among the public, the government restricts mobility – the clear purpose of which is to allay public resentment and anger, rather than to address the root causes of exploitation and abuse. Indeed, when public attention subsides, the government lifts bans and releases restrictions, without making any changes to strengthen protections. Policy restrictions have indeed acted as a façade for the state to escape public criticism and show a commitment to protecting its women migrants going global. Grossman-Thompson (2016, 47) describes this as a "perverse self-perpetuating dynamic" that enables the state to "set the stage for unsafe migration conditions and then rush the stage as the rescuing hero".

Although the Department of Foreign Employment, which retains the main executive authority to regulate the country's labour migration, states that the intent of the bans is to "protect women from risks, including long working hours, sexual violence, physical abuse and economic exploitation", the use of such restrictive policies provides a more structural manifestation of gender inequality. Similarly, such restrictive measures contravene international human rights treaties that Nepal has ratified such as the Convention on the Elimination of All Forms of Discrimination against Women (CEDAW), which obligates signatory states to take steps to eliminate discrimination against women based on gender and to realize women's rights through equal access and opportunities (O'Neil, Fleury, and Foresti 2016). More importantly, the gendered migration policies of the state are clearly contrary to the SDGs and have direct relevance for a number of Goals, namely those pertaining to women's empowerment and equality (e.g. Goals 5, 8, and 10). Further, restrictive labour migration policies are counterproductive in that they limit women's economic opportunities and place women at a greater risk of abuse by giving them less control over their migration experience. Nepali women, hard-hit by poverty and the loss of livelihoods, are pushed to migrate through irregular channels relying on *dalals* (unregistered and unlicensed brokers) who are accountable to no one.

The root of the discriminatory migration policies lies in the state's assumption that often conflates women's labour migration with prostitution/sexual trafficking and that essentializes Nepali women migrants as a homogeneous group of sexual victims. Kharel (2016) observes that women migrants are generally discouraged from migrating due not only to the possibility of exploitation and abuses abroad but also out of the fears of sexual "impurity". O'Neill (2001) contends that such "discourses of national honour" consider women as legitimate objects of state protection, reinforced through the discourse of vulnerability which creates and sustains women's inferiority. The violation of female sexuality is thus the violation of the national sanctity.

The portrayal of women migrants as "minors" and "victims", rather than as citizens with rights, also justifies the "protective" policies of the state on female migration, which in turn forces women migrants to take unauthorized channels for migration (Kharel 2016). Nepal's labour migration policy fails to recognize gender as a key factor and ignores the causes that force women to seek employment abroad, their needs and issues, the tasks they perform, and the challenges they face abroad. The recruitment agencies in Nepal are popularly known as "manpower" agencies – an example of the invisibilization of Nepali women in public policy and discourse. Similarly, although the Foreign Employment Act (2007) and the 2016 guidelines do not discriminate between men and women in "foreign employment", these policies are indifferent to the differential challenges and experiences faced by men and women migrant workers.

Various structural, instrumental, and discursive factors, ranging from the local to the subnational, national, and global, contribute to shaping women migrants' experience at home and abroad. In the first place, the labour migration policy of the sending country has far-reaching ramifications and impacts far beyond its borders. Redressing gaps and discriminatory practices and provisions at the national level of the sending state plays a crucial role in mitigating risks and ensuring safety and rights of women migrant workers abroad. States should therefore formulate and implement appropriate migration policies that are sensitive to the issues of women and women migrants (Hamada 2012). Yet, the Nepali state lacks a clear human rights-based labour migration policy concerning women's cross-border labour mobility, and the absence of such a policy paradigm is a key challenge to ensuring "safe and orderly" migration of women as stipulated in the SDGs. Nepal has signed memoranda of understanding (MoUs) with some GCC countries, but they keep silent on addressing the rights of domestic migrant workers whose occupations are exposed to heightened risks of exploitation and abuse. As of 2017, Nepal has signed memoranda of understanding (MoU) on labour migration with seven countries: Qatar, United Arab Emirates, Republic of Korea, Bahrain, Japan, Israel and Jordan (Nepal Human Rights Commission 2018). However, in most Middle Eastern countries, where Nepali women migrant workers are concentrated, domestic work is not even incorporated into their national labour laws. This happens despite the fact that Guideline 4.5 of the ILO Multilateral Framework on Labour Migration (2005, 12) urges states to ensure that "labour migration policies are gender-sensitive and address the problems and particular abuses women often face in the migration process". The failure to address the issues and lived-realities of women migrant domestic workers represents a significant gap in the bilateral instruments.

An equally daunting challenge, besides tackling the absence of proactive gender-mainstreamed migration policies, is the state's failure to strictly implement the existing national laws pertaining to labour migration. For instance, Nepal's Foreign Employment Act (2007) clearly requires potential migrants to fulfil three legal requirements. First, every migrant must possess all documents duly approved by the Department of Foreign Employment, including a work license. Second, approval measures such as employment contracts specifying wages, hours of work, and other details of employment are required. Third, all Nepali citizens leaving the country for foreign employment are required to depart only from the Nepali airports. Resorting to bribery, the private recruiting agencies frequently evade these rules (Doherty *et al.* 2014).

Nepal issued new Guidelines in 2016 to better govern the recruitment of migrant domestic workers and unleashed its restrictive policy by allowing women over 24 years of age to migrate for domestic work in the Gulf states and Malaysia only with the state-approved recruitment agencies. The Guidelines state that every domestic worker should be entitled to accommodation, health and life insurance, a weekly rest day, 30 days of annual leave, regular contact with their families, and a minimum salary of $300 (Human Rights Watch 2017). The Guidelines, however, while laudable, are difficult to be implemented without negotiating any bilateral agreements with the Gulf countries that continue to implement the *kafala* system. This system can trap domestic workers in abusive conditions because they are not allowed to change jobs without their employer's consent and are punished with imprisonment and deportation if they flee. In March 2017 however, the government imposed a complete ban on Nepali women going to Gulf countries for domestic work under the instructions of the Parliament's International and Labor Relations Committee and this ban still exists (Nepal Human Rights Commission 2018).

Women's agency and empowerment: migration as an alternative survival strategy

Recognizing and enhancing women's freedom and agency by addressing the systemic drivers of gender injustice is vital. States must take intentional measures to address harmful gender-related practices, reduce gender discrimination, and increase women's choices and decision-making power (O'Neil, Fleury, and Foresti 2016). Further, in less developed countries such as Nepal, it is important to recognize that gender discrimination intersects with other factors such as poverty, caste, marital status, ethnicity, disability, and sexuality. The most important question is to empower women in a way that ensures "livelihoods, the ability to enjoy their human rights, a reduction in the unpaid work that hinders the enjoyment of rights, and meaningful participation as actors and leaders in their communities" (Esquivel 2016, 14). For women and girls to exercise their agency freely, structural impediments to gender equality, such as social norms that perpetuate discriminations, need to be challenged and removed. It is important to address the structural and political barriers that stand in the way of equitable governance structures and institutional processes so that women, especially in poor countries such as Nepal, can benefit from migration for employment.

Our contention is that, despite all the structural and institutional barriers, Nepali women's migration for work is a sheer assertion of their agency for survival against all kinds of social, economic, and cultural barriers and vicissitudes. The possibility of migration has given some sections of Nepali women an escape from the patriarchal oppression and empowered them to achieve independence (ILO 2015a). As Massey, Axinn, and Ghimire (2010) contend, labour migration has become an alternative livelihood for most women in Nepal.

The structures of labour migration governance, and the constructed discourse of vulnerability around women's labour mobility lead to the further legitimization and entrenchment of gender discrimination and curtailment of women's human rights. It remains to be seen, however, to what extent the SDGs can push states towards recognizing and changing such deeply embedded structural inequalities. Since the SDGs are understood as universal but nationally differentiated, states are expected to design their own strategies for realizing them. It is not likely that states will recognize their own roles in perpetuating structural violence. Structural violence is exerted indirectly through social, political, and economic systems, and enacted through the social machinery of oppression that shapes suffering (Farmer 2003). Despite their cross-cutting and more nuanced recognition of the complexities of inequality, the SDGs will not inherently tackle such broader structural realities. Indeed, insufficient attention has been given in the SDGs to the role that structural violence plays in the creation of gendered migration experiences.

Gender-based violence, such as domestic violence and forced marriage in the country of origin, drives migration. Sexual violence may also impact migrant women and young girls along migration routes. Thus, an assessment of inequalities and, in particular, of gender-based discrimination and vulnerabilities, is needed to inform policymaking. Second, promoting the opportunities offered by migration for women and girls, such as their education and economic independence, is essential to support their empowerment. Transforming these actions into sustainable opportunities requires robust monitoring systems and data solutions.

Only through adopting a gender-responsive approach can entrenched gender stereotypes and prejudices be challenged. Gender-biased, gender-neutral, and gender-blind labour migration policies have far-reaching implications far beyond their borders. Thus, strengthening national labour migration governance in the sending country is key for effective migration governance (Hamada 2012). Hamada (2012, 57) proposes three interrelated policy goals to be implemented concerning temporary low-skilled migrant workers so that effective migration governance may ensure longer-term benefits for sending countries: "promote employment", "protect and promote wellbeing of migrants," and "maximize developmental impact of labour migration". Nepal has failed to effectively and justifiably harness all of these potentials in the best interests of the individual migrants and the state.

Fleury (2016) contends that migration can improve women's autonomy, self-esteem, and the sense of self-worth in their families and communities. Migration creates "empowerment trade-offs" for women, and these "trade-offs" matter for gender equality and for achieving the SDGs (O'Neil, Fleury, and Foresti 2016). The SDGs provide crucial gender-specific goals, but gender-specific actions are also necessary to empower women and reduce gender inequality (O'Neil, Fleury, and Foresti 2016). Migration creates empowerment trade-offs for individual women, and these trade-offs matter for gender equality and for achieving the SDGs (Foresti and Hagen-Zanker 2017). We know that migrant women contribute to economic development through their vital care labour and through the fiscal contributions they make in offsetting gaps in care systems, in high and low-income countries, in origin and destination countries alike. Additionally, we know that migrant workers contribute more in taxes and social contributions than they receive in individual benefits. This is because women migrant workers tend to pay into employment insurance systems but often are not eligible for maternity leave or parental benefits in the sectors in which they are concentrated. Further, women migrants contribute to development in countries of origin through social as well as financial remittances.

Nepali women's labour migration has not only been instrumental in alleviating poverty (SDG 8) but has also arguably been a means of empowering these women (ILO 2015a). However, the outcomes resulting from women's cross-border labour migration in Nepal, as in any country in the world, cannot be judged solely on the amount of economic remittances women migrants send home. Rather, the evaluation of women's labour migration outcomes must weigh the extent their migration leads to economic empowerment, alongside social development, and gender equality. Only when governance is gender-responsive, when women are enabled to make informed decisions, access legal services, protections, and social networks in both the sending and receiving countries, can there be opportunities for empowerment. Indeed, women should be considered equal stakeholders and contributors to all policy, implementation, and development endeavours, and not merely as a "vulnerable" group in need of the state's patronization.

Women migrant workers – mostly domestic workers – contribute about 50 per cent of migrant workers' remittances, which in Nepal amounts to almost 30 per cent of Gross Domestic Product (UN Women 2013). Women migrant workers also send home a greater share of their remittances (despite earning less) and do so more frequently than their male counterparts (Pérez Orozco, Paiewonsky, and García Domínguez 2008; Petrozziello 2013). They are also more likely to spend monetary remittances on human capital factors such as education and healthcare in their origin countries (IOM 2004; Hennebry, Holliday, and Moniruzzaman 2017). Women migrant workers contribute social remittances in the form of norms, ideas, beliefs, and social capital exchanges across borders that are not usually measured empirically. However, the potential human development outcomes (and women's empowerment) of migration and remittances are constrained by a range of factors including lack of financial inclusion (access to banking, documentation, property, etc.), pay equity gaps, gender discrimination, which are the realities of women's labour migration. In particular, women migrants are concentrated in low-skilled, low-paid and often informal sectors characterized by under regulation, risks of exploitation and violence (including sexual violence), precarity, in addition to barriers to accessing social protection, health care, and social and cultural rights. For many women migrant workers specifically, labour migration means a re-entrenchment of gender discrimination, limited or restrictive pathways into gendered occupations (often those characterized by informality, precarity, and low levels of social protection), deskilling, and a devaluation of care work. So, while migration can create opportunities for women's economic empowerment, it will not inherently do so. Not even the presence of bilateral agreements or government-funded recruitment agencies is sufficient to ensure gender-responsive and development-friendly outcomes of migration. There must be an active and resourced commitment to gender-responsive governance mainstreamed across a range of policy domains (migration, labour, the social sector, etc.).

To facilitate such an approach, the 2030 Agenda can be applied to guide policymakers to mainstream gender into national migration and development policies, as well as to strengthen policy processes to prevent exploitation and abuse of women migrant workers, and to empower them as workers and as women. The SDGs further provide a tool with which women migrants are able to hold policymakers accountable for their labour and human rights. We raised the question in the preceding analysis as to whether the SDGs can aid in tackling more structural and systemic gender inequality – such as addressing restrictive migration governance and exploitative recruitment practices in the context of women's labour migration. Clearly, to come close to that aim, there must be greater international recognition of the role of the global care chain in the economy. Women's labour migration has created a "care drain" in Nepal. The absence of Nepali women in their own households serves to burden younger women in the family, such as daughters or younger sisters who shoulder the remaining domestic work. Similarly, women migrant workers are largely confined to low-paid reproductive or entertainment jobs and this "reflect[s] gendered socio-cultural conditions of varying nature in the sending and receiving countries" (Piper 2003, 724). To break such systemic inequalities, Nepal must invest in training and skills for women, and negotiate bilateral agreements with countries to better protect women migrants' rights.

Conclusion

In migration, gender norms, power relations, and unequal rights between women and men shape women's migration experiences, options, and choices (O'Neil, Fleury, and Foresti 2016). At the family and social levels, gender norms may limit women's migration and prevent them from migrating for fear of "moral corruption" or difficulties in marrying (Shaw 2005). At the national level, states impose restrictive policies to protect women and such an approach not only contravenes women's fundamental human rights for mobility but also falls short of addressing the root causes of the risks women migrants face. Women migrants' bodies are thus the "sites of culturally inscribed and disputed meanings, experiences and feelings" (McDowell 2009, 12).

Gender equality is not just an objective in itself, but it is essential in achieving all development goals. Gender inequality has interconnected economic, political, and social dimensions in which power inequalities and discriminatory norms are embedded. The SDGs do not address the more structural and pervasive nature of gender discrimination and structural violence that is deeply embedded in the existing governance paradigm. The sources and nature of women's subordination and oppression include the feminization of poverty and employment, women's unpaid reproductive and care work, as well as the intersection between vertical and horizontal inequalities based on women's caste, class, marital status, and position in the global economy.

The SDGs also discount the agency of women migrants and misrepresent their diverse local realities. Women's status in society is heavily influenced by their material conditions and by their positions in the national, regional, and global economies. Gender, caste, and race and the social construction of their defining characteristics are interconnected. The post-2015 development framework, while moving past the women-and-development perspective in some respects, does not address the feminization of poverty, women's unpaid labour, human rights violation committed against women, or discrimination based on sexual orientation or gender identity. What would be necessary is an approach that seeks to transform dominant gendered, economic structures of power that would lead to substantive gender equality and women's empowerment.

Restrictive migration policies have not prevented women's labour migration; rather, migration bans have placed women at greater risk of abuse during the migration journey, and limited women's control over their migration decisions (ILO 2015b). These restrictions are specifically aimed at lower-class and lower-middle-class working women who have few options but to migrate for work. While Nepali men have suffered human rights abuses in the Gulf countries, no restrictive migration laws have been put in place to restrict their mobility – and no comparable moral discourse has emerged around their bodies, nor framed male migrants as inherently vulnerable subjects. This way the protectionist and patronizing migration policy aims to consolidate the female "self" that internalizes male superiority and takes patriarchy for granted. As we have demonstrated in the case of Nepali women migrant workers, protectionism has generated transaction costs, promoted wide-spread informality in the recruitment process, impeded the development of a skilled workforce, and harmed the reputation of Nepali women migrant workers in the Middle East. Further, these efforts have undermined women's role in human development in Nepal, despite their continued migration and remittance sending. Gender-mainstreaming Nepal's national labour migration laws and governance mechanisms is necessary not only to ensure Nepali women's fundamental rights to mobility but also to empower them as envisaged in the SDGs.

SDGs call on states to achieve, rather than just promote, gender equality and the empowerment of women and girls. However, gender equality cannot be realized if the status quo of labour migration governance continues unchallenged. The proposed targets include ending violence, eliminating harmful practices, recognizing the value of unpaid care, ensuring that women have full participation – and equal opportunities – in decision-making, and calling for reforms to give women equal access to economic resources. States must recognize that gender inequality is currently both a driver of women's labour migration, and an outcome of the way in which labour migration has been governed. The new post-2015 agenda aims to "leave no one behind", but to make this a reality, the gendered structural violence embedded in labour migration governance must be challenged.

References

Doherty, M., B. Leung, K. Lorenze, and A. Wilmarth. 2014. *Understanding South Asian Labor Migration*. University of Wisconsin-Madison, School of Public Affairs. Workshop in International Public Affairs, Spring. Accessed 10 December 2018. www.lafollette.wisc.edu/images/publications/workshops/2014-migration.pdf.

Esquivel, V. 2016. "Power and the Sustainable Development Goals: A Feminist Analysis". *Gender and Development* 24 (1): 9–23.

Esquivel, V., and C. Sweetman. 2016. "Gender and the Sustainable Development Goals". *Gender and Development* 24 (1): 1–8.

Farmer, P. 2003. *Pathologies of Power: Health, Human Rights, and the New War on the Poor*. Berkeley: University of California Press.

Fleury, A. 2016. "Understanding Women and Migration: A Literature Review". *KNOMAD* Working Paper 8. Accessed 22 November 2018. http://atina.org.rs/sites/default/files/KNOMAD%20Understaning%20Women%20and%20Migration.pdf.

Foresti, M., and J. Hagen-Zanker. 2017. "Introduction". In *Migration and the 2030 Agenda for Sustainable Development*, edited by Overseas Development Institute (ODI), 5–12. Accessed 4 December 2018. www.odi.org/sites/odi.org.uk/files/resource-documents/11751.pdf.

Fukuda-Parr, S. 2016. "From the Millennium Development Goals to the Sustainable Development Goals: Shifts in Purpose, Concept and Politics of Global Goal Setting for Development". *Gender and Development* 24 (1): 43–52.

Grossman-Thompson, B. 2016. "Protection and Paternalism: Narratives of Nepali Women Migrants and the Gender Politics of Discriminatory Labour Migration Policy". *Refuge* 32 (3): 40–48.

Hamada, Y. 2012. "National Governance in International Labour Migration". *Migration and Development* 1 (1): 50–71.

Hennebry, J., J. Holliday, and M. Moniruzzaman. 2017. *At What Cost? Women Migrant Workers, Remittances and Development*. United Nations Entity for Gender Equality and the Empowerment of Women (UN Women).

Human Rights Watch. 2017. "Nepal Events of 2016". *Human Rights Watch*. Accessed 5 December 2018. www.hrw.org/world-report/2017/country-chapters/nepal.

ILO (International Labour Organization). 2005. *The Multilateral Framework on Labour Migration*. Accessed 4 December 2018. www.ilo.org/global/topics/labour-migration/publications/WCMS_178672/lang-en/index.htm.

ILO (International Labour Organization). 2015a. *Analysis of Labour Market and Migration Trends in Nepal*. Kathmandu, Nepal. Accessed 5 December 2018. www.ilo.org/kathmandu/whatwedo/publications/WCMS_407963/lang-en/index.htm.

ILO (International Labour Organization). 2015b. *Realizing a Fair Migration Agenda: Labour flows between Asia and Arab States – Background Paper*. 2–3 December, Kathmandu. Accessed 22 November 2018. www.ilo.org/asia/publications/WCMS_358544/lang-en/index.htm.

IOM (International Organization for Migration). 2004. *Gender, Migration and Remittances*. Accessed 5 December 2018. www.iom.int/sites/default/files/about-iom/Gender-migration-remittances-infosheet.pdf.

Johnson, M., and C. Wilcke. 2015. "Caged in and Breaking Loose: Intimate Labour, the State, and Migrant Domestic Workers in Saudi Arabia and Other Arab States". In *Migrant Encounters: Intimate Labor, the State, and Mobility across Asia*, edited by S. Friedman and P. Mahdavi, 135–159. Philadelphia: University of Pennsylvania Press.

Kharel, A. 2016. "Female Labor Migration and the Restructuring of Migration Discourse: A Study of Female Workers from Chitwan, Nepal". PhD diss., Kansas State University.

Kilkey, M, D. Perrons, and A. Plomien. 2013. *Gender, Migration and Domestic Work: Masculinities, Male Labour and Fathering in the UK and USA (Migration, Diasporas and Citizenship)*. New York: Palgrave Macmillan.

Lan, P. 2008. "Migrant Women's Bodies as Boundary Markers: Reproductive Crisis and Sexual Control in the Ethnic Frontiers of Taiwan". *Signs* 33 (4): 833–861.

Massey, D., S. Axinn, and W. Ghimire 2010. "Environmental Change and Out-Migration: Evidence from Nepal". *Population and Environment* 32 (2): 109–136.

McDowell, L. 2009. *Working Bodies: Interactive Service Employment and Workplace Identities*. Chichester: Wiley-Blackwell.

Nepal Human Rights Commission. 2018. *Trafficking in Persons in Nepal: National Report*. September, Lalitpur, Nepal. Accessed 6 December 2018. www.nhrcnepal.org/nhrc_new/doc/newsletter/NHRC_National_Report_TIP_in_Nepal_September_2018.pdf.

Oksala, J. 2013. "Feminism and Neoliberal Governmentality". *Foucault Studies* 16: 32–53.

O'Neill, T. 2001. "'Selling Girls in Kuwait': Domestic Labour Migration and Trafficking Discourse in Nepal". *Anthropologica* 43 (2): 153–164.

O'Neil, T., A. Fleury, and M. Foresti. 2016. "Women on the Move: Migration, Gender Equality and the 2030 Agenda Sustainable Development". *ODI*, July. Accessed 22 November 2016. www.odi.org/publications/10476-women-move-migration-gender-equality-and-2030-agenda-sustainable-development.

Pérez Orozco, A., D. Paiewonsky and M. García Domínguez. 2008. *Crossing Borders II: Migration and Development from a Gender Perspective.* UN-INSTRAW (now part of UN Women), Santo Domingo, DR.

Peterson, V. S. 2012. "Rethinking Theory: Inequalities, Informalization and Feminist Quandaries". *International Feminist Journal of Politics* 14 (1): 5–35.

Petrozziello, A. J. 2013. *Gender on the Move: Working on the Migration-Development Nexus from a Gender Perspective.* Santo Domingo: UN Women.

Piper, N. 2003. "Feminization of Labor Migration as Violence against Women: International, Regional, and Local Nongovernmental Organization Responses in Asia". *Violence Against Women* 9 (6): 723–745.

Razavi, S. 2016. "The 2030 Agenda: Challenges of Implementation to Attain Gender Equality and Women's Rights". *Gender and Development* 24 (1): 1–17.

Shaw, J. 2005. "Overseas Migration in the Household Economies of Microfinance Clients: Evidence from Sri Lanka". In *Remittances, Microfinance and Development: Building the Link*, edited by J. Shaw, 84–91. Brisbane: The Foundation for Development Cooperation.

Sijapati, B. 2014. *Enhancing Employment-Centric Growth in Nepal: Situational Analysis for the Proposed Employment Policy, Government of Nepal.* ILO Country office for Nepal. Accessed 22 November 2018. www.ilo.org/wcmsp5/groups/public/-asia/-ro-bangkok/-ilo-kathmandu/documents/publication/wcms_245672.pdf.

Sunam, R. 2014. "Marginalised Dalits in International Labour Migration: Reconfiguring Economic and Social Relations in Nepal". *Journal of Ethnic and Migration Studies* 40 (12): 1–19.

UN Women. 2013. Contributions of Migrant Domestic Workers to Sustainable Development. Regional Office for Asia and the Pacific, Bangkok, Thailand, Accessed 8 December 2018. www2.unwomen.org/-/media/field%20office%20eseasia/docs/publications/2013/policy%20paper%20for%20the%20pregfmd%20vi%20high%20level%20regional%20meeting%20on%20migrant%20domestic%20workers.pdf?la=en.

UN (United Nations). 2017. *International Migration Report 2017.* Accessed 4 December 2018. www.un.org/en/development/desa/population/migration/publications/migration port/docs/MigrationReport2017.

UNGA (United Nations General Assembly). 2015. *Transforming Our World: the 2030 Agenda for Sustainable Development.* A/RES/70/1. Accessed 17 November 2018. https://undocs.org/A/RES/70/1.

6 The problem with international migration and sustainable development

Jonathan Crush

> Each of us holds a piece of the migration puzzle, but none has the whole picture. It is time to start putting it together
>
> (Kofi Annan quoted in UNGA 2006b)

Introduction

As the post-2015 global development agenda and Sustainable Development Goals (SDGs) were formulated and debated, there was considerable pressure to include a migration SDG. One of the most notable silences in the Millennium Development Goals (MDGs) was any mention of international migration and there was considerable momentum to rectify this omission in the SDGs. However, the drafters rejected all proposals in this direction and relegated migration to a few general targets scattered among the SDGs. The central conundrum addressed in this chapter is that it is a great deal easier to explain why the problem of international migration led to its omission from the MDGs, than it is to understand why the promise of international migration for development has largely been ignored in the SDGs.

The silence about migration in the MDGs is not because anyone thought the issue unimportant. In the 1980s and 1990s, migration and refugee flows were growing rapidly and becoming more complex, with a notable rise in cross-border migration between countries within the Global South (what later came to be called South–South migration). Vigorous academic debates about the relationship between migration and development were also in full flow. As Portes (2008, 19) later observed:

> the study of international migration and development has been wracked by the controversy between perspectives that see the outflow of people not only as a symptom of underdevelopment but also as a cause of its perpetuation, and those that regard migration both as a short-term safety valve and as a potential long-term instrument for sustained growth.

de Haas (2010) suggests that post-1945 research and policy discourse on the relationship between international migration and development has swung like a pendulum between wild optimism and deep pessimism. The 1990s was a decade of gloom, although it did not begin this way. The Programme of Action of the landmark 1994 Cairo International Conference on Population and Development devoted a whole chapter to international migration and made numerous recommendations on how to make migration a win–win for countries of origin and destination (UNFPA 1994, 82–92). However, there was minimal action on these recommendations in the years that followed despite the best efforts of various UN agencies and international bodies such as the International Organization for Migration (IOM). In a somewhat desperate attempt to move the Cairo Programme of Action forward, the UN and IOM convened a Technical Symposium on Migration and Development in the Hague in 1998. As Castles (1999, 6) observed, the most remarkable feature of this symposium of international organizations and migration experts was that it took place at all. Few governments attended, and none in an official capacity. Castles' (1999, 6) explanation was that international migration is "an area marked by interest conflicts and differing national policy approaches which hinder international cooperation", leading to deep divisions between countries of emigration and immigration who were unable to reach agreement on basic principles. He concluded that "many policy makers still see international migration more as a threat to national security and identity than as an opportunity for cooperation and development. There is no 'international community' with common goals and interests in this area as yet" (Castles 1999, 16).

Conflicts, disagreement, accusations, and counter-accusations over migration (particularly between states in the North and South) both within and outside the UN guaranteed that the issue would be entirely precluded from the MDGs. The IOM later suggested that migration and development was a cross-cutting issue and tried to "retrofit" migration by identifying various ways in which it could contribute to the MDGs and their targets (Usher 2005). The United Nations Population Fund (UNFPA) convened a panel of experts who made essentially the same arguments (UNFPA, 2005). However, once codified, goals and targets are impossible to rewrite and there was very little consideration of international migration in assessments and evaluations of the MDGs, including in the Task Force reports of the UN Millennium Project. Skeldon (2009, 171) later rationalized the omission by asserting that "migration is not a MDG and it is right that migration is not an MDG". His reasoning was that targets for migration were unlikely in the way that targets for reducing poverty, infant or maternal mortality, gender equality, or increasing primary enrolment could be achieved (Skeldon 2009). However, reducing the silence in the MDGs to technocratic questions of measurement overlooks the evidence that there was more inter-state conflict than consensus over migration in the late 1990s and no state wanted migration on the global development agenda. For example, the UN repeatedly attempted to get member states to agree to an international conference on migration. In 1999 it canvassed opinions and found 47 in favour, five partially in favour, and 26 against, with the other 60 per cent failing to even reply to the survey (UNDP 2002).

As this chapter shows, the atmosphere lifted dramatically in the first two decades of the twenty-first century, when international migration was increasingly viewed through the lens of development. Both within the UN system, and outside it, migration no longer seemed to provoke the same anxieties as it had in the 1990s. The forms this realignment took and the institutional and policy changes that resulted are explored in the next section of the chapter. The discursive terrain of "development" provided a basis for talking about migration in new ways and fostered greater inter-state co-operation at the bilateral and multilateral level (Crush 2015). It also led to new institutional arrangements and priorities which fostered a developmentalist approach to migration by many UN agencies, other international organizations (including the IOM, the ILO, and the World Bank), various regional blocs, and individual states.

Redeeming migration through development

After years of inter-state conflict and disagreement over migration and common fears about threats to national sovereignty, the promise of the 1994 Cairo Programme of Action began to be realized after the turn of the century. Within the UN system, migration and development started to achieve greater prominence with the appointment of Kofi Annan as the United Nations Secretary General in 1997. General Assembly resolution 54/212 of 22 December 1999, for example, requested the Secretary General to submit

> a report that will, *inter alia*, summarize the lessons learned, as well as best practices on migration management and policies, from the various activities relating to international migration and development that have been carried out at the regional and interregional levels.
>
> (UNGA 2000, 5)

This was forthcoming in 2002 (UNDP 2002). Meanwhile, General Assembly Resolution 56/203 in 2001 called upon the United Nations system and other organizations "to continue to address the issue of international migration and development and to provide appropriate support for processes and activities on international migration" (UNGA 2001, 4).

In 2002, the UN convened a system-wide Coordination Meeting on International Migration. But Annan's efforts to constitute a UN commission on international migration were unsuccessful, stymied by countries in the Global North. His response was to take the commission idea outside the UN system and an independent Global Commission on International Migration (co-chaired by high-profile individuals from Sweden and South Africa) was appointed in 2003 with Annan's blessing. By the time the Commission finished its global hearings and issued its report in 2005, it had been overtaken by events within the UN (GCIM 2005). In particular, UN General Assembly Resolution 58/208 of 2003 resolved to convene a high-level dialogue on international migration and development during its 61st session in 2006.

The first ever UN High Level Dialogue (HLD) on International Migration and Development was inaugurated by the Secretary General's Report which asserted that "international migration constitutes an ideal means of promoting co-development, that is, the coordinated or concerted improvement of economic conditions in both areas of origin and areas of destination based on the complementarities between them" (UNGA 2006a). Most state inputs at the HLD echoed these sentiments although there was a divergence of views on whether the UN should create a dedicated migration agency, with states in the Global North resolutely opposed. Despite, or perhaps because of, this setback, many individual UN agencies began to incorporate migration and development into their programming after 2006. In 2009, the flagship UN Human Development Report was devoted entirely to the theme of human mobility and development (UNDP 2009) and the UNDP headed a Joint Migration and Development Initiative (JMDI) with the United Nations High Commissioner for Refugees (UNHCR), UNFPA, UN Women, the International Labour Organization (ILO) and the IOM between 2008 and 2012 under the rubric "migration4development".[1] The JMDI's primary stated purpose was to provide policymakers and practitioners with evidence-based recommendations in the field of migration and development.

Further impetus to programming on migration and development within the UN system was provided by the formation, by the Secretary General, of the Global Migration Group (GMG) in 2006. The GMG grew to comprise the heads of 22 UN agencies plus the World Bank, the World Health Organization (WHO), and the IOM. Its stated aim was "to promote the wider application of all relevant international and regional instruments and norms relating to migration, and to encourage the adoption of more coherent, comprehensive and better coordinated approaches to the issue of international migration".[2] In 2010, the GMG established an inter-agency *Working Group on Mainstreaming Migration into National Development Strategies* to develop tools and resources to enhance the GMG *Handbook on Mainstreaming Migration into Development Planning* (GMG 2010). Other objectives included the promotion of a human development lens in migration and development research, analysis and policy development at the country level, and fostering a common approach by international agencies to mainstream migration into development planning at the country level. Further GMG activities under the migration and development remit include convening an International Conference on Harnessing Migration, Remittances and Diaspora Contributions for Financing Sustainable Development in 2015, the formation of a migration and data research working group, making inputs to the Global Forum on Migration and Development and the launch, in late 2017, of the *Handbook for Improving the Production and Use of Migration Data for Development* (GMG 2017).

In 2013, the UN General Assembly convened a second High-Level Dialogue on Migration and Development with the theme of "Making Migration Work". The GMG made a strong plea to the HLD delegates to integrate migrants into all facets of UN development work: "our key message for the High-Level

Dialogue on International Migration and Development and beyond is that migration, undeniably, is an enabler of inclusive and sustainable development for migrants and societies alike" (GMG 2013, 1). The General Assembly adopted a Declaration on Migration and Development with 34 resolutions which included a number of key commitments (UNGA 2013):

- International migration should be addressed in a coherent, comprehensive and balanced manner, integrating development with due regard for social, economic and environmental dimensions and respecting human rights.
- To acknowledge the important contribution made by migrants and migration to development in countries of origin, transit and destination, as well as the complex interrelationship between migration and development.
- To work towards an effective and inclusive agenda on international migration that integrates development and respects human rights by improving the performance of existing institutions and frameworks, as well as partnering more effectively with all stakeholders involved in international migration and development at the regional and global levels.
- To strengthen synergies between international migration and development at the global, regional, and national levels.
- To recognize the efforts made by the international community in addressing relevant aspects of international migration and development, through different initiatives, both within the United Nations system and in other processes, particularly the Global Forum on Migration and Development and regional processes, as well as in drawing on the expertise of the International Organization for Migration and other member agencies of the Global Migration Group.
- To acknowledge that the United Nations system can benefit from the discussions and outcomes of the Global Forum on Migration and Development, in order to maximize the benefits of international migration for development.
- To acknowledge the important role that migrants play as partners in the development of countries of origin, transit and destination and recognize the need to improve public perceptions of migrants and migration.

The post-2000 surge of global interest in the relationship between international migration and development was also evident outside the UN system. The emphasis of the inter-governmental IOM shifted towards a more overtly developmentalist agenda under the directorship of William Lacey Swing. This was reflected in a host of global, regional, and country-specific programming initiatives, the high-profile annual migration report (see IOM 2013a), and the annual inter-governmental International Dialogue on Migration.[3] The World Bank, with its work on remittances and Global Knowledge Partnership on Migration and Development (KNOMAD) initiative also took a lead role in the generation and sponsorship of migration research for economic development (World Bank 2016).

At the international level, the Global Forum on Migration and Development (GFMD) (which emerged out of the UN HLD 2006 as an alternative to a new agency within the UN system) has become the major international meeting place for the exchange of information, ideas and policy options on migration and development. The GFMD has now met ten times since its launch in Belgium in 2007, alternating between venues in the Global North and South. The Forum has its critics (particularly relating to its relative neglect of migrant rights see Martin and Abella 2009; Rother 2009; Wee, Vanyoro, and Jinnah 2018). But as a non-binding inter-governmental forum, there is little doubt that it has been a significant catalyst for inter-governmental policy debates and discussions about international migration as a lever for development in countries of migrant origin through remittances, diaspora engagement, and temporary migration (Newland 2012; Omelaniuk 2016).

International migration and development has also been a growing focus of attention in interactions within and between regional blocs of states. The European Union's Global Approach to Migration and Mobility (GAMM), for example, has as one of its four core objectives "maximising the development impact of migration and mobility" (EC 2005; Hampshire 2016). Or again, the African Union (AU) committed itself and member states to integrate migration and development in the 2006 Migration Policy Framework (AU 2006a), the Revised Migration Policy Framework and Plan of Action (AU 2016), and the AU Common Position on Migration and Development (AU 2006b). Achieng (2012) has reviewed the priorities of multi-state Regional Consultative Processes on Migration (RCPs) and highlighted GFMD proposals for bringing the migration and development nexus onto their agendas (GFMD 2010). Inter-regional processes (particularly those with a North–South axis) already have migration and development as a key modality through various kinds of partnership. For example, the Joint Africa–EU Declaration on Migration and Development, the EU–AU Valetta Accord and the EU–ACP Migration Partnership all place migration and development issues at their centre (Crush 2015; Klavert 2011; Trouille 2016).

While many states remain fixated on international migration control and threats to territorial sovereignty, the mainstreaming of migration into development discourse has made progress at this scale as well. At the 2013 HLD, several states described plans to incorporate migration into development planning. Jamaica, for example, alerted the General Assembly to its National Policy and Plan of Action on Migration and Development. And Lesotho drew attention to its new Draft Comprehensive Migration and Development Policy. Perhaps the Philippines has gone furthest to date with a wide range of migration and development-related programmes and initiatives (Nicolas 2016). One of the key policy mechanisms linking international migration and development at the country level is engagement with the diaspora (Chikanda, Crush, and Walton-Roberts 2016). Gamlen (2014) found that over half of all states in the UN now have a government institution with a diaspora engagement brief. States have also been sharing best practice strategies at international fora, including the

GFMD and the first ever Diaspora Ministerial Conference in 2013, attended by as many as 143 governments (IOM 2013b). Viewing international migration through the lens of development therefore overcame most, if not all, of the barriers that shut it out of the MDGs.

A sprinkling of migration issues

In 2013, the late Peter Sutherland, Special Representative of the UN Secretary-General on Migration from 2006 to 2017, wrote a piece entitled "Migration is Development" and observed:

> To succeed, the post-2015 agenda must break the original mold. It must be grounded in a fuller narrative about how development occurs – a narrative that accounts for complex issues such as migration. Otherwise, the global development agenda could lose its relevance, and thus its grip on stakeholders.
>
> (Sutherland 2013)

One of the agenda items at the 2013 UN HLD envisioned the international community defining a common set of targets and indicators to monitor the implementation of measures aimed at enhancing the benefits and addressing the challenges of international migration (UNGA 2013). At the HLD itself, a roundtable on "Assessing the effects of international migration on sustainable development and identifying relevant priorities in view of the preparation of the post-2015 development framework" made a strong argument for migration, asserted that the time has come to make migration an integral part of the next global development agenda. Lazcko and Lönnback (2013) provided many concrete suggestions for a migration-related goal, targets, and indicators on behalf of the IOM.

Amongst the 36 recommendations adopted by the General Assembly were two relating to the post-2015 sustainable development agenda:

- To acknowledge the important contribution of migration in realizing the Millennium Development Goals, and recognize that human mobility is a key factor for sustainable development which should be adequately considered in the elaboration of the post-2015 development agenda (UNGA, 2013, 2).
- To call upon all relevant bodies, agencies, funds and programmes of the United Nations system, other relevant intergovernmental, regional and subregional organizations … to strengthen their collaboration and cooperation to better and fully address the issue of international migration and development, in order to adopt a coherent, comprehensive and coordinated approach, and to consider migration issues in their contributions to the preparatory process that will establish the post-2015 development agenda (UNGA, 2013, 5).

Thus, on the eve of the formulation of the post-2015 development agenda and the SDGs, the General Assembly agreed on a set of recommendations centring the role of migration in sustainable development and agreeing that it should be incorporated into the goals.

In tandem with these HLD resolutions, there was a vigorous lobbying effort to ensure that international migration had a prominent place in the Sustainable Development Goals, what Sutherland later referred to as "three long years of advocacy".[4] Bakewell (2015) characterizes this effort as follows:

> There has been a growing clamour to bring the topic into the mainstream of development policy and practice. The main cheerleaders have been the International Organization for Migration (IOM) along with interested migration programme units in donor governments, UN agencies, the World Bank and an array of civil society organisations, in particular migrant rights and diaspora groups. In particular, there was great concern to ensure that the SDGs should take account of migration in some way, something the MDGs had failed to do. This stimulated a barrage of initiatives, workshops, debates and dialogues to find ways to insert migration into the SDGs.

The IOM fronted much of the advocacy "barrage", working in public and behind the scenes to try and get greater recognition for a migration SDG or, failing that, the systematic integration of migration into the post-2015 development agenda (Laczko and Lönnback 2013). The Global Migration Group also called to have migration explicitly recognized in the post-2015 global development agenda:

> We hope to have a new reference text for global development, one which builds on and enhances the Millennium Development Goals (MDGs). That text, we hope, will have remedied the MDGs' silence on migration issues. The new text will reflect the realization that migration is not exclusively a South–North issue, but contributes to equitable, inclusive and sustainable development in all countries and regions. A few years from now, we hope to be implementing a new "partnership for migration" as part of a renewed global partnership for development.
>
> (GMG 2013, 3)

Although Sutherland labelled the presence of migration-related issues in a handful of SDG targets a "triumph", it was a hollow victory. The lobbying effort largely failed to have the desired impact on the Open Working Group (OWG) responsible for drafting the SDGs. Nor did the OWG give effect to the resolutions of the 2013 HLD. There is no migration SDG when arguably there could have been, given the decade or more of high-profile global momentum and programming on migration and development documented above. As Fendrich (2014) points out, migration's admission to the post-2015 agenda was "discrete" "instead of a systematic attempt to link migration to broader development goals, one should rather speak of a 'sprinkling' of migration issues". Newland (2017, 3) further remarks that "the largely implicit presence of migration in the Sustainable Development Goals shows that international migration still plays a relatively minor role in international development thinking".

The major concession to the migration and development lobby was a paragraph in "Transforming our World: The 2030 Agenda for Sustainable Development" which referenced "the positive contribution of migrants for inclusive growth and sustainable development" and noted that "international migration is a multidimensional reality of major relevance for the development of countries of origin" (UNGA 2015, 8). However, there is a major disjuncture between these sentiments and the actual goals and targets. None of the 17 SDGs and only one of the 167 targets refer explicitly to an aspect of migration and development, i.e. the commitment tacked onto the end of the SDG 10 targets to "reduce to less than 3% the transaction costs of migrant remittances and eliminate remittance corridors with costs higher than 5%" (UNGA 2015, 21). Three targets refer to the need to eliminate various forms of trafficking (5.2, 8.7, and 16.2) which is hardly central to the migration and development agenda. Two refer to migrants in passing: 8.8 refers to the protection of labour rights and safe working environments (including for migrants) and 17.18 says that development data should be disaggregated by migratory status amongst many other things.

The only other target with explicit migration content, and an obvious concession to the IOM and its advocacy of "managed migration" (Ashutosh and Mountz 2011) is 10.7: "to facilitate orderly, safe, regular and responsible migration and mobility of people, including through implementation of planned and well-managed migration policies" (UNGA 2015, 21). As Peter Sutherland points out, "unfortunately, States tend to have quite different conceptions of what 'well-managed migration' means in practice. Some would like it to mean more migration; others, no migration at all" (UNGA 2017, 6). "Transforming Our World" and the SDGs make no reference to other key elements of the migration and development policy debate including diasporas and diaspora engagement, skills migration and the brain drain, the "triple win" of temporary migration, return migration, addressing the "root causes" of migration through development financing, migrant and refugee entrepreneurship, and so on. The representation of migration and development in the SDGs is therefore spotty, with more emphasis on migration control than the development implications of migration.

Somewhat disingenuously, organizations in search of institutional relevance to the post-2015 development agenda have subsequently asserted that migration is centrally reflected in the SDGs. For example, the IOM Director General argued in his opening address to the 106th IOM Council Meeting in November 2015 that "with the SDGs, migration is now seen as an issue to act upon to enhance sustainable development".[5] The IOM (2018) has also claimed that:

> Migration is included for the first time in the global development framework, recognizing well-managed migration's integral role in and immense

contribution to sustainable development. The SDGs are now driving policy planning and implementation across borders and across sectors, recognizing the interlinkages between migration and development and the fundamental contributions of migrants.

The IOM organized two International Migration Dialogue events in 2016 with the theme "Follow-up and Review of Migration in the Sustainable Development Goals (SDGs)" (IOM 2017). The organization's reinvention of the SDG agenda is an understandable effort to stake an ongoing claim to institutional relevance (and development finance). While (mis)representations of the SDGs may convince some that the migration-development nexus will be carried forward, this is not grounded in any sensible reading of the evidence and contradicts the obvious marginalization of migration and development in the OWG process. Some scholars have nevertheless argued that even the shadowy presence of migration in the SDGs may produce some positive outcomes. Piper (2017, 236), for example, argues that the SDGs "will be an important avenue to protect the rights of migrant workers".

Conclusion

Explaining why migration and development is marginal to the 2030 Agenda, despite a decade or more of high-profile activity within and outside the UN, is beyond the scope of this chapter. However, several preliminary arguments can be advanced. First, it might be that the reasons that kept migration out of the MDGs had a hangover effect, particularly when it came to reconciling the different interests and perspectives of predominantly migrant-sending and migrant-receiving countries. However, one of the major achievements of the period 2000 has been to stake out areas of common interest between states by viewing migration through the lens of development.

Second, following Skeldon (2009, 171) it could be that migration is not an SDG and it is right that migration is not an SDG. In other words, the task of getting states to agree on the wording of a goal and drawing up quantifiable targets and indicators for migration would have been difficult. However, this essentially technocratic argument is undercut by a decade of reflection on appropriate migration and development indicators. It would not have taken a great deal of imagination to formulate a set of targets and indicators under a general goal of making migration work for sustainable development (Laczko and Lönnback 2013).

Third, it could be that the SDGs process was loaded against migration from the start. An examination of the recommendations of the High-Level Panel of Eminent Persons on the Post-2015 Development Agenda, and those aspects of the Open Working Group process in the public domain, show that migration and development did not feature strongly in any of the SDG preparatory documents and discussions (UN 2013[6]). This may in part be attributed to conceptualizations of sustainable development and accompanying neglect of migration in the UN's parallel sustainable development process, which culminated in the 2012 UN Conference on Sustainable Development in Rio de Janeiro (Rio + 20).

The first OWG draft of proposed goals and targets in June 2014 hardly referenced migration at all, which suggests that the migration lobby did have some minor success in achieving some added targets in the year that followed. However, despite the growing profile of migration and development within the UN system, the main lobbyists for a migration SDG were from outside the system and undoubtedly had much less power to advocate as a result. The "outsider" status of the IOM effectively changed in 2016 when it became a UN agency. It is tempting to speculate that if the new "insider" status had come a little earlier, there would have been more than a sprinkling of migration in the SDGs.

Finally, the acceptance of migration as a quintessentially development issue with and outside the UN may have been undercut by the more pressing and immediate migration challenges of the day – crisis migration, the "mixed migration" of refugees and economic migrants, "mass migration" to Europe from Africa and the Middle East, and the increased securitization of migration by states (UNGA 2016). Suliman (2017) argues that the inclusion of migration in the SDGs cannot serve as the basis for the resolution of the contemporary migration crisis. Certainly, the SDGs provide little guidance on how these issues should be addressed, although it could explain their undue emphasis on trafficking. There now seems to be growing realization that processes and mechanisms other than the SDGs are likely to be more effective and necessary to move a global migration and development agenda forward within the UN (Newland 2017; UNGA 2017).

Notes

1 For more on this initiative, please see www.migration4development.org/en/content/about-jmdi. Accessed 29 November 2018.
2 Extracted from www.globalmigrationgroup.org. Accessed 29 November 2018.
3 See www.iom.int/international-dialogue-migration. Accessed 29 November 2018.
4 See the Message of Peter Sutherland, Special Representative of the UN Secretary General for International Migration and Development, at the 2016 International Dialogue on Migration, 29 February–1 March 2016, New York. www.migration4 development.org/en/content/about-jmdi Accessed 29 November 2018.
5 See the 106th IOM COUNCIL – 26th November 2015, Geneva, Switzerland. High-level Panel Discussion on Sustainable Development Goals and IOM Migration Governance Framework. www.youtube.com/watch?v=wsvWH-P-OsM. Accessed 29 November 2018.
6 See also the Open Working Group on Sustainable Development Goal's webpage https://sustainabledevelopment.un.org/owg.html Accessed 29 November 2018.

References

Achieng, M. 2012. "Regional and Inter-Regional Processes: Advancing the Discourse and Action on Migration and Development". In *Global Perspectives on Migration and Development*, edited by I. Omelaniuk, 187–120. Dordrecht: Springer.

Ashutosh, I., and A. Mountz. 2011. "Migration Management for the Benefit of Whom? Interrogating the Work of the International Organization for Migration." *Citizenship Studies* 15 (1): 21–38.

AU (African Union). 2006a. *The Migration Policy Framework for Africa*. Accessed 29 November 2018. www.unhcr.org/protection/migration/4d5258ab9/african-union-migration-policy-framework-africa.html.

AU (African Union). 2006b. *African Common Position on Migration and Development*. Accessed 29 November 2018. www.un.org/en/africa/osaa/pdf/au/cap_migrationand dev_2006.pdf.

AU (African Union). 2016. *Revised Migration Policy Framework for Africa and Plan of Action (2018–2027)*. Accessed 29 November 2018. https://au.int/sites/default/files/newsevents/workingdocuments/32718-wd-english_revised_au_migration_policy_framework_for_africa.pdf.

Bakewell, O. 2015. "Migration Makes the Sustainable Development Goals Agenda: Time to Celebrate?" *Oxford Martin School*, 11 December. Accessed 29 November 2018. www.oxfordmartin.ox.ac.uk/opinion/view/315.

Castles, S. 1999. "International Migration and the Global Agenda: Reflections on the 1998 UN Technical Symposium". *International Migration* 37: 5–19.

Chikanda, A., J. Crush, and M. Walton-Roberts, eds. 2016. *Diasporas, Development and Governance*. Dordrecht: Springer.

Crush, J. 2015. "The EU–ACP Migration and Development Relationship". *Migration and Development* 4: 39–54.

de Haas, H. 2010. "Migration and Development: A Theoretical Perspective". *International Migration Review* 44: 227–264.

EC (European Commission). 2005. "Global Approach to Migration and Mobility". *European Commission: Migration and Home Affairs*. Accessed 29 November 2018. https://ec.europa.eu/home-affairs/what-we-do/policies/international-affairs/global-approach-to-migration_en.

Fendrich, P. 2014. "Included or Excluded? Migration in the Post-2015: An Assessment of the Proposal of the Open Working Group on Sustainable Development Goals". *Poverty-Wellbeing*, August. Accessed 29 November 2018. www.shareweb.ch/site/Poverty-Wellbeing/current-poverty-issues/sdgs-agenda2030/included-or-excluded-migration-in-the-post-2015.

Gamlen, A. 2014. "Diaspora Institutions and Diaspora Governance". *International Migration Review* 48: S1890–S1917.

GCIM (Global Commission on International Migration). 2005. *Migration in an Inter-Connected World: New Directions for Action*. Geneva: Global Commission on International Migration.

GFMD (Global Forum on Migration and Development). 2010. *How Can Regional Consultative Processes on Migration (RCPs) and Inter-Regional Fora (IRF) Best Include the Migration and Development Nexus*. Background Paper for the Global Forum on Migration and Development, Puerto Vallarta.

GMG (Global Migration Group). 2010. *Mainstreaming Migration into Development Planning: A Handbook for Policy-Makers and Practitioners*. Accessed 29 November 2018. www.globalmigrationgroup.org/system/files/uploads/UNCT_Corner/theme7/main streamingmigration.pdf.

GMG (Global Migration Group). 2013. "UN General Assembly High-Level Dialogue on International Migration and Development 2013: Statement by the Global Migration Group 'International Migration and Development'". Accessed 29 November 2018.

www.globalmigrationgroup.org/system/files/uploads/news/GMG-statement-for-2013-High-Level-Dialogue.pdf.

GMG (Global Migration Group). 2017. *Handbook for Improving the Production and Use of Migration Data for Development*. Accessed 29 November 2018. www.globalmigration group.org/system/files/Handbook_for_Improving_the_Production_and_Use_of_Migration_Data_for_Development.pdf.

Hampshire, J. 2016. "Speaking with One Voice? The European Union's Global Approach to Migration and Mobility and the Limits of International Migration Cooperation". *Journal of Ethnic and Migration Studies* 42: 571–586.

IOM (International Organization for Migration) 2013a. *World Migration Report 2013: Migrant Well-Being and Development*. Geneva: International Organization for Migration.

IOM (International Organization for Migration). 2013b. *Diasporas and Development: Building Societies and States*. Geneva: International Organization for Migration.

IOM (International Organization for Migration). 2017. *Follow-Up and Review of Migration in the Sustainable Development Goals*. Geneva: International Organization for Migration.

IOM (International Organization for Migration). 2018. "2030 Agenda for Sustainable Development". Accessed 29 November 2018. https://unofficeny.iom.int/2030-agenda-sustainable-development.

Klavert, H. 2011. "African Union Frameworks for Migration: Current Issues and Questions for the Future". Discussion Paper No. 108, European Centre for Development Policy Management, Maastricht.

Laczko, F., and L. Lönnback, eds. 2013. *Migration and the United Nations Post-2015 Development Agenda*. Geneva: IOM.

Martin, P., and M. Abella 2009. "Migration and Development: The Elusive Link at the GFMD". *International Migration Review* 43: 439–443.

Newland, K. 2012. "The GFMD and the Governance of International Migration". In *Global Perspectives on Migration and Development*, edited by I. Omelaniuk, 227–240. Dordrecht: Springer.

Newland, K. 2017. "The Global Compact for Migration: How Does Development Fit In?" *Towards a Global Compact for Migration: A Development Perspective*, Issue 1. Washington DC: Migration Policy Institute.

Nicolas, I. 2016. "Engaging the Global Filipino Diaspora: Achieving Inclusive Growth". In *Diasporas, Development and Governance*, edited by A. Chikanda, J. Crush, and M. Walton-Roberts, 33–47. Dordrecht: Springer.

Omelaniuk, I. 2016. "The Global Forum on Migration and Development and Diaspora Engagement". In *Diasporas, Development and Governance*, edited by A. Chikanda, J. Crush, and M. Walton-Roberts, 19–32. Dordrecht: Springer.

Piper, N. 2017. "Migration and the SDGs". *Global Social Policy* 17: 231–238.

Portes, A. 2008. "Migration and Development: A Conceptual Review of the Evidence". In *Migration and Development: Perspectives from the South*, edited by S. Castles and R. Delgado Wise, 17–42. Geneva: IOM.

Rother, S. 2009. "The Allure of Non-Binding Talks: The Global Forum on Migration and Development". *Development and Cooperation* 36: 331–333.

Skeldon, R. 2009. "Migration Policies and the Millennium Development Goals". In *A Progressive Agenda for Global Action*, 170–180. London: Policy Network.

Suliman, S. 2017. "Migration and Development After 2015". *Globalizations* 14: 415–431.

Sutherland, P. 2013. "Migration is Development". *Project Syndicate*, 15 March. Accessed 29 November 2018. www.project-syndicate.org/commentary/migrants-and-the-post-2015-global-development-agenda-by-peter-sutherland?barrier=accesspaylog.

Trouille, J. 2016. "From Mass Migrations to Sustainable Development. Re-thinking the EU–Africa Partnership as a Win–Win Co-Development Strategy". In *Jean Monnet Seminar on Migrations*. 22–23 February, Tunis, Tunisia. Luxembourg: European Commission Directorate General for Education and Culture, Publications Office of the European Union.

UN. 2013. *A New Global Partnership: Eradicate Poverty and Transform Economies Through Sustainable Development. Report of the High-Level Panel of Eminent Persons on the Post-2015 Development Agenda*. Accessed 29 November 2018. www.post2015hlp.org/the-report/.

UNDP (United Nations Development Programme). 2002. *International Migration Report 2002*. New York: United Nations.

UNDP (United Nations Development Programme). 2009. *Overcoming Barriers: Human Mobility and Development, Human Development Report 2009*. New York.

UNFPA (United Nations Population Fund). 1994. *Programme of Action Adopted at the International Conference on Population and Development*, Cairo, 5–13 September 1994. Accessed 29 November 2018. www.unfpa.org/sites/default/files/event-pdf/PoA_en.pdf.

UNFPA (United Nations Population Fund). 2005. *International Migration and the Millennium Development Goals: Selected Papers of the UNFPA Expert Group Meeting*. Accessed 29 November 2018. www.unfpa.org/sites/default/files/resource-pdf/migration_report_2005.pdf.

UNGA (United Nations General Assembly). 2000. *International Migration and Development*. A/RES/54/212, 22 December. Accessed 29 November 2018. www.iom.int/jahia/webdav/shared/shared/mainsite/policy_and_research/un/54/A_RES_54_212_en.pdf.

UNGA (United Nations General Assembly). 2001. *International Migration and Development*. A/RES/56/203, 21 February. Accessed 29 November 2018. www.iom.int/jahia/webdav/shared/shared/mainsite/policy_and_research/un/54/A_RES_54_212_en.pdf.

UNGA (United Nations General Assembly). 2006a. *Report of the Secretary-General on International Migration and Development*. A/60/871, 60th session. Accessed 29 November 2018. www.un.org/esa/population/migration/hld/.

UNGA (United Nations General Assembly). 2006b. UN General Assembly takes up issue of migration, 14–15 September in New York. Accessed 20 February 2019. www.un.org/migration/medialert.html.

UNGA (United Nations General Assembly). 2013. *Declaration of the High-level Dialogue on International Migration and Development*. A/68/L.5, 1 October. Accessed 29 November 2018. https://documents-dds-ny.un.org/doc/UNDOC/GEN/N13/439/69/PDF/N13 43969.pdf.

UNGA (United Nations General Assembly). 2015. *Transforming Our World: the 2030 Agenda for Sustainable Development*. A/RES/70/1. Accessed 17 November 2018. https://undocs.org/A/RES/70/1

UNGA (United Nations General Assembly). 2016. *In Safety and Dignity: Addressing Large Movements of Refugees and Migrants: Report of the Secretary-General*. A/70/59, 70th session. Accessed 29 November 2018. https://refugeesmigrants.un.org/2016-secretary-generals-report.

UNGA (United Nations General Assembly). 2017. *Report of the Special Representative of the Secretary-General on Migration*. A/71/728. Accessed 29 November 2018. https://

documents-dds-ny.un.org/doc/UNDOC/GEN/N17/002/18/PDF/N1700218.pdf?Open Element.

Usher, E. 2005. *The Millennium Development Goals and Migration*. Migration Research Series No. 20. Geneva: IOM.

Wee, K., and K. Vanyoro, and Z. Jinnah. 2018. "Repoliticizing International Migration Narratives? Critical Reflections on the Civil Society Days of the Global Forum on Migration and Development". *Globalizations* 15 (6): 795–808.

World Bank. 2016. "Migration and Development: A Role for the World Bank Group". Accessed 29 November 2018. http://documents.worldbank.org/curated/en/6903814726 77671445/pdf/108105-BR-PUBLIC-SecM2016-0242-2.pdf.

7 SDGs and climate change adaptation in Asian megacities

Synergies and opportunities for transformation

Idowu Ajibade and Michael Egge

Introduction

The 2018 Intergovernmental Panel on Climate Change (IPCC) report provides a sobering overview of the state of the world's climate. Human activities are estimated to have caused approximately 1.0 degree Celsius of global warming above pre-industrial levels and warming is likely to reach 1.5 degrees between 2030 and 2050 (IPCC 2018). On the development end, millions of people across the world live in abject poverty with limited access to clean water, energy sources, adequate food, quality housing, and sanitation. Climate change is likely to exacerbate these problems (see Schweizer, Chapter 9). Under a 1.5 degree increase in global temperature, several Asian cities may experience intensifying rainfall, typhoons, heatwaves, and flooding with consequences for water-related disasters, increased disease vectors, major disruptions in food and agricultural systems, and erosion of infrastructural and livelihood stability (Hallegatte *et al.* 2013; Pal and Eltahir 2016; Williams *et al.* 2016). These threats call for revisiting adaptation and development challenges in cities.

Currently, 137 million people in Asia reside in flood plain areas and this number is expected to rise to 200 million by 2030 due to rapid population growth and urbanization processes (ADB 2012; Neumann *et al.* 2015). As cities grow, the population of people in informal settlements is also expected to increase (Barrett, Horne, and Fien 2016; UN Habitat 2016). On the one hand, informal settlements represent the adaptability, innovation, and resilience of city dwellers, but on the other hand, they portray the failure of cities to address wealth inequality, housing shortages, and socially unjust urban development patterns (Grimm *et al.* 2008). Shifting demographic trends together with climatic-fluctuations will put significant pressure on the urban landscape and on the resource economy base that supports urban populations. Environmental degradation, intensifying consumption, and globalization will add to these existing pressures in Asia. Furthermore, multinational companies operating in the Asian growth economies will also be exposed to spiralling environmental risks over the coming decades (World Bank 2012). This daunting scenario reinforces the need for a radical change in the adaptation-development nexus of Asian cities, particularly if they are to transition towards sustainability in an equitable and inclusive manner within an acceptable risk-threshold.

We argue that stronger co-ordination and synergies between the Sustainable Development Goals (SDGs) and climate change adaptation offer new opportunities for a robust response to existing development problems and to new climatic risks facing many Southeast Asian megacities. At present, a growing momentum exists to unleash the transformative potential of adaptation on the one hand, and sustainable development on the other. However, the extent to which current efforts on both issues are able to reduce inequalities, improve living conditions, ensure economic growth, and guarantee safe, resilient, and sustainable urbanization is unclear. To explore this issue, we examine 30 city-scale climate adaptation strategies (plans, policies, projects, and programs) in three Southeast Asian cities: Manila, Jakarta, and Bangkok. All three cities have active adaptation plans and Disaster Risk Reduction (DRR) programmes implemented by various government agencies and non-governmental organizations (NGOs), including the United Nations (UN), World Bank, Oxfam, Save the Children, and World Vision, among others. Many of the adaptation strategies we analysed already commenced before the SDGs were established. Since adaptation is an on-going process, it will be crucial to harmonize them with the SDGs.

We recognize that adaptation is a continuous process and often difficult to measure (Pelling 2010), therefore we do not assess adaptations for their success or failure. Rather, we explore the extent to which these strategies complement or conflict with SDGs 10 and 11. We also identify the underlying objectives informing these adaptation strategies and explore to what extent they contribute to long-term environmental and social sustainability. While the results of this adaptation are specific to Manila, Bangkok, and Jakarta, they, in part, provide relevant lessons that are applicable to other megacities in the Global South.

What follows is a brief history of the SDGs with particular focus on SDGs 10 and 11. We then explore the complementary relationship between the SDGs and climate change adaptation. The third section discusses the case studies and findings. We explore the similarities and differences in each city's approach to adaptation and sustainability. We conclude the chapter by reiterating the importance of transforming current adaptation patterns as well as building greater synergies between the SDGs and on-going adaptation efforts in cities.

MDGs to SDGs

The SDGs replace the previously established Millennium Development Goals (MDGs) (2000–2015). There are 17 SDGs with 169 targets, covering a broad range of sustainable development issues far more expansive than the MDGs. While significant progress was achieved on the MDGs, disparities across and within countries remained. Globally, the number of people living in extreme poverty was reduced by half and the proportion of undernourished people decreased from 23.2 per cent in 1990 to 14.9 per cent in 2012 (United Nations 2015). Between 1990 and 2012, the population living on less than US$1.25 per day decreased from 1.7 billion people to 569 million (United Nations 2015). China's successful poverty reduction strategy accounted for most of the world's gain over the last 20 years (Permanyer 2013). East Asia also saw a significant drop in extreme poverty rate from 61 per cent in 1990 to 4 per cent in 2015 (United Nations 2015). In contrast, many South Asian countries, including India, failed to meet the MDGs (United Nations 2015) and more than 40 per cent of the population in sub-Saharan Africa remain in poverty (United Nations 2015).

The East and Southeast Asia experience on MDGs is indicative of the positive impacts of setting a global agenda on development and investing resources in such an agenda. However, despite the progress achieved, a number of old problems persist, while new ones have emerged. Growth of the region's economy occurred in tandem with growth in urban population. In Southeast Asia, 41.8 percent (172 million people) of the population live in urban areas and this number is projected to increase to 50 per cent by 2025 (UN Habitat 2013). Economic growth has also meant an increase in income inequality, urban density, and higher disparity in living conditions. Close to 75 million people in the region live on less than $3.10 a day, most of whom live in informal settlements (World Bank 2017). Residents of slums often lack the security of land tenure and access to basic infrastructure, such as adequate water supply, sanitation, and proper removal and management of waste disposal. In addition, emissions of carbon dioxide are on the rise just as land use change, pollution levels, weather-related hazards, and human vulnerability to climatic and environmental risk are increasing.

The SDGs framework provides an opportunity to re-examine the role of cities in engendering sustainable development in the context of a changing climate. SDGs 10 and 11 are particularly relevant. These two goals focus on reducing disparities and promoting sustainable living. SDG 11 specifically calls for making cities and human settlement safe, inclusive, resilient, and sustainable by 2030, while SDG 10 seeks to reduce inequality within and among countries. To achieve SDG 10, countries are expected to adopt and implement fiscal and social protection policies that progressively reduce inequality while also taking measurable actions to promote inclusion of all people, independent of age, sex, disability, race, ethnicity, religion, economic or other status (United Nations 2018). SDG 11 builds on SDG 10 by encouraging cities to create plans to upgrade slums, increase access to safe and affordable housing, improve air quality, and foster participatory urban planning and management. Goal 11 also emphasizes adopting policies aimed at engendering resource efficiency, mitigation and adaptation to climate change, and disaster resilience in line with the Sendai Framework for Disaster Risk Reduction 2015–2030 (United Nations 2018). An integrated approach to achieving SDGs 10 and 11 therefore requires building stronger synergies between disaster risk reduction, development, and climate change adaptation. This kind of synergy calls for buy-in from a variety of sectors and actors as well as planning processes that incorporate these goals as part of a broader sustainability plan for cities.

Globally the discourse on adaptation and development is intertwined with resilience and sustainability. By sustainability, we refer to a state of affairs that ensures a better quality of life for all, now and into the future, in a just and equitable manner, while living within the limits of supporting ecosystems and reducing ecological impacts (WCED 1987; Agyeman, Bullard, and Evans 2003). Some scholars see the SDGs and climate change adaptation as two distinct issues aimed at the broader goal of sustainability. Others have highlighted the need to reframe adaptation to better align it with visions of a sustainably developed world (Hackmann *et al.* 2013; Wise *et al.* 2014; Eriksen, Nightingale, and Eakin 2015; Ajibade 2017), where adaptation together with mitigation are considered fundamental to avoiding disruptions in the Earth's climate system and to enhancing human capacity to manage and benefit from perceived changes (Parry *et al.* 2007). Climate change adaptation creates new and, in many ways, unprecedented challenges for cities, especially where they have to respond to high magnitude events such as the 2017 Houston floods and 2017 Hurricane Maria that set Puerto Rico's development back by several years. Such events often exceed current adaptive capacities of cities in terms of available resources, equitable response, and risk mitigation (Garschagen and Kraas 2011). Similarly, achieving the SDGs in a just, inclusive, and sustainable manner will require resources and capacity building which some cities may not have. Addressing these two issues in a synergistic manner therefore creates new opportunities for cities to re-organize their future priorities in ways that could lead to success on both ends.

Study context: Manila, Jakarta, and Bangkok

Metro-Manila, Jakarta, and Bangkok are three major urban centres in Southeast Asia. They rank among the top cities in the region that are vulnerable to tropical storms, sea-level rise, land subsidence, and climate change-related flooding. The topography of these cities plays a role in their physical vulnerability. Metro-Manila, the capital region of the Philippines, lies on the flat alluvial and deltaic plains draining the Pasig River and Laguna de Bay (Department of Environment and Natural Resources 2018). It is composed of 16 cities and has a land mass of approximately 636 km² (Ragragio 2003). The population of the metro-area is 12.8 million people with a density of 20,000 persons per km² (Philippines Statistics Authority 2016). A large portion of the city is built on the coastal margin, including the reclaimed areas in Manila Bay. The climate is characterized by relatively high temperatures and frequent rainfall. Several parts of the metro area are flood-prone. Heavy rains, land subsidence, and rising sea levels contribute to flooding, especially in informal settlements where an estimated four million people live (Habitat for Humanity 2017).

Jakarta is a low-lying coastal city, built on a swampy plain on the island of Java in Indonesia. The city has a population of over 10 million and a density of 14,000 persons per km² (World Population Review 2018). The land mass of the city is 661.48 km² extending to the Jakarta Bay on the northwest coast of the

Island. The tropical climate provides stable, warm temperature year around, while the monsoon season lasts throughout the winter, with the wet season itself stretching into fall and spring. Jakarta is sinking faster than any other city in the world; significant parts of the city lie below sea level and experience heavy rains which make flooding a perennial challenge that affects thousands of people. For example, from 2002 to 2014 four major floods occurred in the city displacing about one million people, most of whom live in low-income crowded neighbourhoods that lack public services such as transportation, water supply, sanitation, and waste management (Padawangi and Douglass 2015; Amri *et al.* 2016). Flood vulnerability in the city is exacerbated by population growth, rapid urbanization, changing land use, and land subsidence (Sagala *et al.* 2013; Abidin *et al.* 2011). As the greater Jakarta mega-urban region approaches the 30 million population mark, the sources of flooding may become ever more complex through combinations of climate change and human transformations of the urban landscape (Padawangi and Douglass 2015).

Similar to Manila and Jakarta, Bangkok has a relatively high urban population of about eight million in the city proper, but the metro region is home to 15.3 million people living within 2,590 km^2 (Thanvisitthpon, Shrestha, and Pal 2018). Bangkok city lies in the Chao Phraya delta where it empties into the Gulf of Thailand (Losiri *et al.* 2016). The city experiences a warm, tropical climate with a summer monsoon season which results in seasonal flooding on the low-lying former swamplands that make up the current city (Thanvisitthpon, Shrestha, and Pal 2018). In the 2011 monsoon season, severe flooding paralyzed large parts of the city, triggering greater awareness about climate change and its impact on development and urban sustainability. Thailand has been one of the widely cited development success stories with sustained economic growth and impressive poverty reduction since the MDGs. However, inequality and inadequate access to housing pose significant challenges. In Bangkok, the share of urban population living in slums is 25 per cent (ADB 2014) indicating the city has some work to do to meet SDGs 10 and 11.

Wealth inequality is far worse in the Philippines and Indonesia. In the Philippines, the 40 richest families accounted for 76 per cent of the country's gross domestic product growth; and two of the wealthiest people in the country are worth 6 per cent of the nation's entire economy (France-Presse 2013; Sahakian and Dunand 2014). In Indonesia, the four richest men accounted for the combined wealth of the poorest 100 million people (Oxfam 2017). Market fundamentalism in both Indonesia and the Philippines has allowed the richest to capture the benefits of strong economic growth and concentration of land ownership. Growing inequality will undermine the fight against poverty and could threaten social cohesion. As climate change worsens, wealth inequalities in these cities will too, thus widening the gap between the haves and the have nots. Efforts to build resilience and sustainable development in these cities therefore must take into consideration inequality and poverty rates.

In this chapter, we explore the following questions: are current adaptation strategies in cities compatible or in conflict with the SDGs? What are the barriers to the progressive realization of SDGs 10 and 11 in megacities? And how can these barriers be addressed in ways that allow for transformation towards long-term urban sustainability?

Adaptation strategies and compatibility with the SDGs

Of the 30 city-level adaptation strategies analysed in this study, 10 were directly compatible with SDGs 10 and 11 while others conflicted with both goals.

In Metro-Manila, we identified four adaptations that are strongly compatible with the SDGs: the Metro-Manila Green Print 2030; the Makati City Disaster Vulnerability Plan; the Metro Bus Rapid Transit Project; and the Valenzuela City Partners for Resilience Program. The Metro-Manila Green Print aims to achieve the following: upgrade informal settlement citywide, create affordable housing, reduce greenhouse gas (GHG) emissions, address disaster vulnerability across sectors and groups, and promote efficient land use and sustainable economic growth. There are also efforts to prevent human settlement in high-risk land and flood-prone areas (Zhang et al. 2014). The Makati Disaster Vulnerability Plan has a similar aim – it seeks to reduce GHG emissions and pollution from solid waste while increasing green spaces and efficient land use through urban densification. Improved access to social services and universal health care and immunization for all residents are also part of this vulnerability reduction plan (Prasad et al. 2009). Both plans further outline efforts to reduce inequality while ensuring the safety and resilience of all people in the city, thus contributing to the progressive realization of SDGs 10 and 11.

Adding to the above plans are special projects which includes the Metro-Manila Bus Rapid Transit project. This aims to increase services to low-income residents, while decreasing their overall transit cost and improving safety and accessibility for children and the elderly (World Bank 2017). The project will also decrease traffic congestion and increase usage of public transit, thereby reducing GHG emissions (World Bank 2017). Another is the Valenzuela City Partners for Resilience Program which focuses on new development practices that improve the resilience of communities and increase environmental amenities, such as the expansion of green spaces, trees, and parks. The program further incorporates scientific forecasts into the development of worst-case scenarios and contingency planning alongside community drills. Workshops and projects organized at community levels have engaged and empowered three-quarters of the Valenzuela population in hazard response plans, greening, and social learning. In particular, community members learn to identify who will do what, where, when, and how, based on climate-informed forecasts (Mateo and Lagdameo 2015). These different projects and plans in Metro-Manila can be further leverage towards achieving the SDGs by 2030.

In the case of Jakarta, the Jakarta Participatory Mapping of Disaster Risk and Preparedness project is compatible with the SDGs. Its overarching goal is to educate the public and local government on ways to adapt to climate change and proactively reduce risks (World Bank 2013). The project engages multiple groups such as government officials, practitioners, students, and citizens on hazard mappings and seeks to evenly disseminate disaster preparedness information and contingency planning. Through this project, risk data were collected for over 2,668 neighbourhoods and 6,000 buildings and critical infrastructure including schools, hospitals, places of worships, and flood-prone areas using Geographic Information System (GIS) software and Open Street Maps (World Bank 2013). This allowed for populations not often captured in data analysis to be better served based on their needs and identities. Stakeholders were also given access to data for emergency planning and for development and poverty reduction projects (World Bank 2013), thus boosting local understanding on adaptation-development nexus.

Another compatible strategy is the Jakarta Education Initiative. It focuses on educating all school children in Jakarta and nationwide on disaster risk reduction and on cultural and behavioural changes that can enhance resilience. In Jakarta, a post-disaster assessment of the 2013 flood reported that more than 70,000 students from 251 primary schools could not access their school for three to four weeks due to flooding (Amri *et al.* 2016). The DRR Education Initiative therefore addresses such concerns by improving students' knowledge on flood prevention and response. While several aspects of the initiative have been hailed as successful, with children and teachers demonstrating knowledge on disaster preparedness and response, there are also concerns about the lack of dedicated personnel and budget to sustain and to scale up the program (Amri *et al.* 2016).

The Jakarta Master Plan 2010–2030, with a fulfilment deadline similar to the SDGs, is perhaps one of the most controversial adaptation plans in the city. At first glance, the plan appeared to be compatible with the SDGs. It expressly focuses on using spatial planning and urban revitalization to address longstanding flooding and housing problems. The key strategies for accomplishing this include: the development of a new sustainable city, land reclamation, resettlement of vulnerable groups, macro infrastructural investments, and financing for sea wall construction (Jakarta Master Plan 2014). This plan also indicates the planting of a mangrove forest and wetlands to mitigate floods and the construction of vertical housing in newly reclaimed areas in North Jakarta, i.e. Kelapa and Penjaringan. It specifically stated that the development of the reclaimed land in the northern coastal areas is designed for the upper-middle class (Jakarta Master Plan 2014, 74) yet the most vulnerable are the low-income class.

Another problem with this plan is the failure to provide structural codes or guidelines for the safety of new high-rise buildings, thus potentially creating new risks for people in an environment prone to land subsidence. Second, low income communities in slum areas such as the *Kampungs* will see little to no improvement in their housing conditions while their resettlement from such areas is deemed necessary for "making room for the river" and for dike construction along the coastlines (Jakarta Master Plan 2014, 53). The public have therefore shown little support for the plan due to its top-down planning process and its infrastructure-oriented rather than people-centred goals (Kusumawijaya and

Sutanudja 2010). If followed through, the Master Plan may lead to Jakarta becoming a highly segregated city with increased social tensions and environmental vulnerabilities. Since its release, several citizen groups have resisted the plan, doubting its credibility and legitimacy (Kusumawijaya and Sutanudja 2010), but there is little indication that the government will change or revise this plan. The 2014 version analysed in this study showed little changes from the 2010 version.

In the case of Bangkok, Thailand's Climate Change Master Plan (2015–2050), combined with the National Economic and Social Development Plan (NESDP) (2012–2016), reflects an adaptation-development policy that is mutually reinforcing. While the former seeks to achieve sustainable low-carbon growth and climate change resilience by 2050, the latter aims to change the development paradigm towards an environmentally friendly economy while expanding green areas in cities. It also focuses on improving adaptation capacity and resilience by developing a plan to deal with the negative consequences of climate change and international policies aimed at the problem (Nachmany *et al.* 2015). Bangkok's Master Plan on Climate Change (2013–2023) is also being developed for the metropolitan area to achieve these goals (JICA 2018).

The Bangkok Rapid Recovery Plan created following the 2011 Thai floods is designed for surprise and change. It incorporates development and governance concerns to improve the city's resilience. The Plan specifically addresses the need for recovery in low-income communities by focusing on jobs for community members of all classes while supporting capacity building to improve people's living conditions. It further aims to increase social understandings of risk and resilience through education; community involvement in decision making; promotion of flexible institutions; better watershed management to control floods; efficient land use and increases in the amount of protected lands. This plan, while compatible with SDGs 10 and 11 requires a follow-through on implementation.

After the 2011 Thai flood, the widespread use of social media to increase flood resilience emerged as a new adaptation strategy in Bangkok. This strategy is not state co-ordinated per se but evolved "organically" as individuals and groups realized they could supplement the lack of an expensive gauging station using social media as an early warning tool. Poor co-ordination, lack of early warning systems, institutional traps, and poor response from government were, in part, responsible for the high damages and loss during the 2011 floods (Lebel, Manuta, and Garden 2011; Marks and Thomalla 2017). Since then, social media has allowed people in different communities to successfully leverage and share information in a timely fashion about flood levels, directions of flows, coping strategies, and temporary relocation (Allaire 2016). The strategy has fostered solidarity and community bonding on a wider scale where different communities sought to help one another to reduce risk. Success has been reported in terms of effective disaster preparedness and reduction in losses in subsequent years (Allaire 2016). By preventing and reducing losses from flood events, this adaptation strategy contributes to social stability, reduced poverty, and increased urban resilience, and is therefore compatible with SDGs 10 and 11. However, it is worth noting that social media is mainly applicable in communities where people have access to the internet and those that are literate may benefit more.

These different strategies (plans, projects and policies) from the three megacities demonstrate that it is possible to build resilient and sustainable urban communities through an integrated approach that combines sustainable development, disaster risk reduction, ecosystem restoration and protection, and climate change adaptation.

Adaptation as barriers to the SDGs and long-term urban sustainability

While a number of adaptation strategies in all three cities contribute to the progressive realization of SDGs 10 and 11, some conflicted with them and could further undermine social equity, resilience, and urban sustainability. To discuss this conflict, we draw on four themes that emerged from our analysis: *treating symptoms instead of root causes; focusing on structural adaptation without attention to everyday problems; high financial costs, elite capture, and corruption; and risk redistribution through adaptation.*

In all three cities, most of the adaptation plans and projects examined did not address the key political economy drivers of disaster vulnerability, poverty, and unsustainability, but rather treated the symptoms as the cause. A number of projects also did not address the complexities and redistributive impacts of specific adaptation projects on the urban poor. For example, the Makati Disaster Vulnerability Plan emphasizes increasing urban densification programs as a solution to reducing GHG emissions in the long term, but this also means having a high number of people live in small areas in an already congested city. By implication, there will be increased pressure on land, water, and energy in the area, including a high demand for air conditioning with consequences for high urban island heat. In the event of a natural or human-induced disaster such as fire, the number of human casualties in densely populated buildings in the city will be high, and impacts could be worse for the poor due to limited assets for recovery.

Furthermore, in all three cities, there is high capital investment in structural adaptation and new city construction that has little impacts on the everyday life of the urban poor. For example, in Bangkok the construction of flood walls to reduce risk to large-scale enterprises and wealthy inner-city areas redistributed risk to unprotected areas and to vulnerable populations (Marks and Thomalla 2017). No significant changes in land use have occurred since the 2011 flood, while the construction of new roads, dykes, and the forcible resettlement of low-income communities have taken place (Yarina 2018). Despite the resilience plans and policies adopted, the city has not been rebuilt better and more equitably; and implementation of these plans have been slow or yet to materialize (Marks and Thomalla 2017).

Similar problems exist in Jakarta, where the government plans to build a giant sea wall called the "Great Garuda". The 15-mile-long sea wall structure is expected to strengthen existing onshore embankment of the Jakarta Bay and also protect the new city being developed on the artificial island reclaimed from the ocean (Jakarta Master Plan 2014; Win 2017). This project has the endorsement of President Joko Widodo, who expressly stated his support for the 2014 version of the Jakarta Master Plan that included details of the construction of the new city as a key adaptation strategy. Jakarta's Great Garuda, estimated to cost $40 billion backed by aid from the Dutch government and a Dutch-led consortium, attempts to change the socio-hydrology of the bay through changes to the water flow and flood patterns (Sherwell 2016). In terms of implications, the project is likely to trigger increased urbanization in flood-prone areas, redirect floods to places not protected by the sea wall, and destroy local fisheries and ecosystem resources that local communities depend upon for their livelihoods, thus undermining SDGs 10 and 11.

Since many adaptations of top priority are capital intensive, there are concerns about elite capture, corruption, and poor fiscal practices and misallocation of project funds. The Manila Green Prints and Bus Rapid Transit (BRT) plan will cost $64.6 million; $40.7 million loaned from the World Bank and $23.9 million loaned from the Clean Technology Fund (World Bank 2017). In Manila, there are several instances where the government was found to misallocate project funds (Pido 2017). The history of indebtedness and unpaid World Bank loans in the Philippines suggests this can have grave consequences on the country's development trajectory, leading to the imposition of austerity measures to recuperate donor funds (Cruz and Repetto 1992). In Indonesia, elite capture, bureaucratic corruption, and bribery are high while transparency is low (Transparency International 2018). Nearly half of the businesses in the land management sector report expecting to give gifts in order to obtain construction permits; property rights are also inadequately protected; and laws and regulation are vaguely worded, resulting in uncertainty and rent-seeking (GAN Integrity 2018). These problems create distrust of government and undermine credibility and legitimacy of capital-intensive adaptation projects such as the Great Garuda, thus serving as barriers to the realization of the SDGs. Responding to a variety of protests against the Great Garuda, the Government of Jakarta in December 2017 announced the cancellation of the project (Anya and Wijaya 2017).

Beyond planned adaptation at the city levels, individuals and communities also engage in autonomous adaptation of their own. Some involve robust collaboration among community members (such as the social media resilience strategy) while others are crude strategies such as unco-ordinated housing improvement in certain parts of the community. The latter, at best, serves as a temporary solution to prevent damages to properties and livelihoods but could generate problems in the long term. In Bangkok, studies show that residents engaged in autonomous adaptation to floods may reduce their own risk, but also shift these risks toward others (Limthongsakul, Nitivattananon, and Arifwidodo 2017). Furthermore, recent development in many low-income communities has increased flooding problems for some and not for others, since residents adapt on their own terms without regard to possible consequences on those living in shanties (Limthongsakul, Nitivattananon, and Arifwidodo 2017). In other words, overall risk is not addressed but shifted within geographic spaces and across groups. Such poor planning puts additional burden on the most vulnerable and economically marginalized who may not have the resources to improve their living conditions. These types of adaptation can also create distrust and anger among community members, thus reducing the potential for them to act together towards achieving resilience and long-term sustainability.

Transforming adaptation to better achieve the SDGs

While certain adaptation strategies in these megacities are flawed, there are opportunities to transform and better align them with the SDGs. Three things, in particular, need attention: revisiting planning principles, using proven development strategies to create systemic change, and transforming the governance of adaptation and the institutions involved.

To make adaptation align better with the SDGs, master plans and major adaptation projects have to be revisited with a new lens. This means using specific guidelines in planning and evaluating adaptation strategies for their likelihood to contribute to transformation towards sustainability. Such principles have been identified in the transformation and adaptation literature to include increases in: participatory visioning; social learning; social equity in planning and outcome; cultural and behavioural shift; vulnerability reduction; ecosystem protection; resource efficiency; sustainable economic growth; and flexible and accountable governance (Park *et al.* 2012; Asara *et al.* 2015; Feola 2015; Eriksen, Nightingale, and Eakin 2015; Fook 2017). Several of these principles were also stated in the IPCC (2018) report as key attributes of adaptation and mitigation pathways that can engender sustainable development. We evaluated all 30 adaptation strategies discussed in this chapter using these principles and we found that the majority of the projects focused on risk reduction, economic growth, and, to some extent, social learning and resource efficiency – all of which are compatible with all three city's Master Plans and overall adaptation strategies. However, the limited attention to concerns around social equity, participatory planning and visioning, accountability, and cultural and behavioural shifts poses significant problems for long-term sustainability and resilience. To achieve the SDGs, policymakers therefore have to reconsider incorporating these transformational principles into their adaptation-development plans.

As discussed above, the top priority adaptations in all three cities are tailored mostly to serve the wealthy. Achieving SDGs 10 and 11 in cities will therefore require systemic changes in the political economy by avoiding elite capture of development and adaptation projects as well as building on pro-poor development programs with proven results. An example of pro-poor development is the Baan Mankong slum upgrade programme in Bangkok. The programme, initiated in 2003, empowers poor communities to take ownership of their own housing development. Instead of delivering housing units to individual poor families, the government allows existing slum communities to form co-ops in order to access government-sponsored subsidized loans to upgrade their settlements (Norford and Virsilas 2016). This program also supports networks of poor communities to survey and map all poor and informal settlements across the city and develop plans to comprehensively upgrade them. Residents work with experts from local governments, NGOs, and academia to execute the project, but the planning, scoping, budgets, infrastructural upgrade, and negotiation for secure land tenure are all community-driven. This type of project is not without its pitfalls; however, it emphasizes participatory visioning and engages the poor as agents of change in the context of the sustainable development and adaptation, thus putting them in the driver's seat to determine their future (Norford and Virsilas 2016). A similar slum upgrade plan has been adopted in Manila under the name "Securing the Safety of Informal Settlers' Families in Metro Manila" (SSISF) (Balgos 2016).

Another example of a people-oriented change from below is the World Bank-sponsored second and third urban poverty projects in Jakarta. The projects focused on improving urban poor livelihoods and empowering them to act to improve drainage and water quality in their communities. The project appraisal showed evidence of little mismanagement and there was a high buy-in from community members who demonstrated this through land donations (World Bank 2015). The final report suggests the project reduced the risk of water pollution, flooding, and potential health hazards through support for home repairs and increased access to health clinics. Community's participation in the project allowed for social learning and bonding. These types of projects can potentially contribute to resilience in poor communities, while allowing for the progressive realization of SDGs 10 and 11.

One of the limitations to adaptation-development synergy is the lack of clear policies and regulatory regimes on how both issues should be addressed, who should address what, and when. Our review of the adaptation policies across all three cities revealed a lack of holistic legislation and inter-agency co-ordination on adaptation and development. Thailand, for example, has development, risk mitigation, and adaptation, scattered in a plenitude of policies and plans, but has not passed a concrete legislation on adaptation (Nachmany *et al.* 2015). A danger in this governance model is that adaptation and development needs get lost in bureaucratic processes, while limited resources end up wasted in the process. In view of achieving the SDGs, state authorities have to harmonize diverse institutions, programmes, and agencies responsible for adaptation and development to ensure greater co-ordination and robust response on both issues. To do this, governance and institutional reform may be necessary in addition to establishing new monitoring agencies and metrics to measure progress.

The institutional harmonization required to implement a coherent framework on adaptation and the SDGs comes with significant financial implications. For example, the expansion and revitalization of the current institutions, cost of staffing, training, monitoring and project review, and program funding are expensive. Due to limited resources and competing budgetary priorities,

SDGs 10 and 11 may run into implementation problems, especially in mega-cities with high population and high development needs. Renewed attention is therefore required on how international and national government can support and finance plans and projects that will lead to realizing the SDGs at the city level. Honest and open discussion on how city governments can mobilize necessary public and private sector financing and capacity to boost inclusive, resilient, and sustainable programs must form part of the conversation.

Conclusion

This study explored the compatibility and conflict between existing adaptation strategies and sustainable development goals in three megacities focusing mainly on Goal 10 (reduce inequality) and Goal 11 (building safe, resilient, and sustainable cities). We also examined how specific adaptation strategies can serve as a barrier to the SDGs and to long-term urban sustainability. We found ten city-level adaptation strategies to be compatible with the SDGs; however, specific aspects of compatible strategies also conflicted with the SDGs. Strategies with the highest investment and overwhelming government support were not compatible with the SDGs as they were driven by economic and risk reduction incentives rather than equity and social sustainability. Treatment of symptoms rather than the root causes of vulnerability and uneven development was prevalent in all three cities. This contradiction demands attention from policy-makers, adaptation planners, and development practitioners. Revision of master plans and reform of institutions and governance structures are crucial to achieving SDGs 10 and 11. City planners and communities together with adaptation experts and development practitioners have to collectively respond to this challenge through socially and environmentally meaningful initiatives that foster inclusion and resilience. Furthermore, increased institutional synergy on development and adaptation together with harmonization of policies and plans to overturn existing gaps will be required as cities look towards realizing the SDGs by 2030.

References

Abidin, H. Z., H. Andreas, I. Gumilar, Y. Fukuda, Y. E. Pohan, and T. Deguchi. 2011. "Land Subsidence of Jakarta (Indonesia) and Its Relation with Urban Development". *Natural Hazards* 59 (3): 1753.

ADB (Asian Development Bank). 2012. *Green Urbanization in Asia: Key Indicators for Asia and the Pacific*. Accessed 26 November 2018. www.wilsoncenter.org/sites/default/files/Green%20Urbanization%20in%20Asia%20Special%20Chapter.pdf#page=28.

ADB (Asian Development Bank). 2014. *Urban Poverty in Asia*. Accessed 20 October 2018. www.adb.org/sites/default/files/publication/59778/urban-poverty-asia.pdf.

Agyeman, J., R. D. Bullard, and B. Evans, eds. 2003. *Just Sustainabilities: Development in an Unequal World*. London: Earthscan.

Ajibade, I. 2017. "Can a Future City Enhance Urban Resilience and Sustainability? A Political Ecology Analysis of Eko Atlantic City, Nigeria". *International Journal of Disaster Risk Reduction* 26: 85–92.

Ajibade, I. and E. A. Adams. 2019 "Planning Principles and Assessment of Transformational Adaptation: Towards a Refined Ethical Approach". *Climate and Development* (online): 1–13.

Allaire, M. C. 2016. "Disaster Loss and Social Media: Can Online Information Increase Flood Resilience?". *Water Resources Research* 52 (9): 7408–7423.

Amri, A., D. K. Bird, K. Ronan, K. Haynes, and B. Towers. 2016. "Disaster Risk Reduction Education in Indonesia: Challenges and Recommendations for Scaling Up". *Natural Hazards and Earth System Sciences* 17: 595–612.

Anya, A., and C. Wijaya. 2017. "Govt Cancels Great Garuda Seawall". *The Jakarta Post*. Accessed 10 November 2018. www.thejakartapost.com/news/2017/12/11/govt-cancels-great-garuda-seawall.html.

Asara, V., I. Otero, F. Demaria, and E. Corbera. 2015. "Socially Sustainable Degrowth as a Social–Ecological Transformation: Repoliticizing Sustainability". *Sustainability Science* 10 (3): 375–384.

Balgos, B. C. 2016. "Securing the Safety of Informal Settler Families Along Waterways in Metro Manila, Philippines: Government-Civil Society Organisation Partnership". In *Disaster Governance in Urbanising Asia*, edited by M. A. Miller, and M. Douglass, 177–193. Singapore: Springer.

Barrett, B. F., R. Horne, and J. Fien. 2016. "The Ethical City: A Rationale for an Urgent New Urban Agenda". *Sustainability* 8 (11): 1–14.

Cruz, W., and R. Repetto. 1992. *The Environmental Effects of Stabilization and Structural Adjustment Programs: the Philippines Case*. No. 338.952 C7. Washington, DC: World Resources Institute.

Department of Environment and Natural Resources. 2018. *National Capital Region Profile*. Accessed 2 November 2018. http://ncr.denr.gov.ph/index.php/about-us/regional-profile.

Eriksen, S. H., A. J. Nightingale, and H. Eakin. 2015. "Reframing Adaptation: The Political Nature of Climate Change Adaptation". *Global Environmental Change* 35: 523–533.

Feola, G. 2015. "Societal Transformation in Response to Global Environmental Change: A Review of Emerging Concepts". *Ambio* 44 (5): 376–390.

Fook, T. 2017. "Transformational Processes for Community-Focused Adaptation and Social Change: A Synthesis". *Climate and Development* 9 (1): 5–21.

France-Presse, A. 2013. "Philippines' Elite Swallow Country's New Wealth". *Inquirer. Net*, 3 March. Accessed 2 November 2018. https://business.inquirer.net/110413/philippines-elite-swallow-countrys-new-wealth.

GAN Integrity. 2018. *Indonesia Corruption Report*. Accessed 26 November 2018. www.business-anti-corruption.com/country-profiles/indonesia/.

Garschagen, M., and F. Kraas. 2011. "Urban Climate Change Adaptation in the Context of Transformation: Lessons from Vietnam". In *Resilient Cities: Responding to Peak Oil and Climate Change*, edited by P. Newman, T. Beattley, and H. Boyer, 131–139. Dordrecht: Springer.

Grimm, N. B., S. H. Faeth, N. E. Golubiewski, C. L. Redman, J. Wu, X. Bai, and J. M. Briggs. 2008. "Global Change and the Ecology of Cities". *Science* 319 (5864): 756–760.

Habitat for Humanity. 2017. "Upgrading Slums in the Philippines: The Need for Social Housing". *Habitat for Humanity Great Britain*, 10 October. Accessed 2 November 2018.

www.habitatforhumanity.org.uk/blog/2017/10/upgrading-slums-philippines-need-social-housing/.

Hackmann, H., S. Moser, K. O'Brien, J. D. Sachs, M. Leach, K. Raworth, J. Rockström, *et al.* 2013. "Part I: The Complexity and Urgency of Global Environmental Change and Social Transformation". In *World Social Science Report: Changing Global Environments*, 65–70. Paris: UNESCO Publishing.

Hallegatte, S., C. Green, R. J. Nicholls, and J. Corfee-Morlot. 2013. "Future Flood Losses in Major Coastal Cities". *Nature Climate Change* 3 (9): 802–806.

IPCC (Intergovernmental Panel on Climate Change). 2018. *Global Warming of 1.50C: An IPCC Special Report on the Impacts of Global Warming of 1. 50C above Pre-Industrial Levels and Related Global Green Gas Emission Pathways, in the Context of Strengthening the Global Response to the Threat of Climate Change, Sustainable Development, and Efforts to Eradicate Poverty.* Accessed 26 November 2018. www.ipcc.ch/report/sr15/.

Jakarta Master Plan. 2014. *The Master Plan of the National Capital Integrated Coastal Development.* Accessed 2 November 2018. www.bureauanl.nl/files/MP-final-NCICD-LR.pdf.

JICA (Japan International Cooperation Agency). 2018. *Project for Bangkok Master Plan on Climate Change 2013–2023 (BANGKOK).* Accessed 2 November 2018. www.mcrit. com/climamollet/index.php?option=com_content&view=article&id=119:project-for-bangkok-master-plan-on-climate-change-2013-2023&catid=79:2015-11-02-12-24-04 &Itemid=71.

Kusumawijaya, M., and E. Sutanudja. 2010. "Why Do We Need to Redo Jakarta Spatial Masterplan?". *The Jakarta Post*, 6 February. Accessed 2 November 2018. www.thejak artapost.com/news/2010/02/06/why-do-we-need-redo-jakarta-spatial-masterplan.html.

Limthongsakul, S., V. Nitivattananon, and S. D. Arifwidodo. 2017. "Localized Flooding and Autonomous Adaptation in Peri-Urban Bangkok". *Environment and Urbanization* 29 (1): 51–68.

Lebel, L., J. B. Manuta, and P. Garden. 2011. "Institutional Traps and Vulnerability to Changes in Climate and Flood Regimes in Thailand". *Regional Environmental Change* 11 (1): 45–58.

Losiri, C., M. Nagai, S. Ninsawat, and R. P. Shrestha. 2016. "Modelling Urban Expansion in Bangkok Metropolitan Region Using Demographic–Economic Data through Cellular Automata-Markov Chain and Multi-Layer Perceptron-Markov Chain Models". *Sustainability* 8 (7): 1–23.

Marks, D., and F. Thomalla. 2017. "Responses to the 2011 Floods in Central Thailand: Perpetuating the Vulnerability of Small and Medium Enterprises?". *Natural Hazards* 87 (2): 1147–1165.

Mateo, M., and D. Lagdameo. 2015. *Building Resilience to Climate Change Locally: The Case of Valenzuela City, Metro Manila March.* Climate and Development Knowledge Network. Accessed 2 November 2018. https://cdkn.org/wp-content/uploads/2015/03/Valenzuela-Inside-Story.pdf.

Nachmany, M., S. Fankhauser, J. Davidová, N. Kingsmill, T. Landesman, H. Roppongi, P. Schleifer, *et al.* 2015. *Climate Change Legislation in Thailand: An Excerpt from the 2015 Global Climate Legislation Study: A Review of Climate Change Legislation in 99 Countries.* Accessed 26 November 2018. www.lse.ac.uk/GranthamInstitute/wp-content/uploads/2015/05/THAILAND.pdf.

Neumann, B., A. T. Vafeidis, J. Zimmermann, and R. J. Nicholls. 2015. "Future Coastal Population Growth and Exposure to Sea-Level Rise and Coastal Flooding: A Global Assessment". *PLOS ONE* 10 (3): 1–34.

Norford, E., and T. Virsilas. 2016. "What Can We Learn from Thailand's Inclusive Approach to Upgrading Informal Settlements?". *The City Fix*, 12 May. Accessed 1 November 2018. http://thecityfix.com/blog/thailands-inclusive-upgrading-informal-settlements-terra-virsilas-emily-norford/.

Oxfam. 2017. "Inequality in Indonesia: Millions Kept in Poverty". *Oxfam International*. Accessed 15 October 2018. www.oxfam.org/en/indonesia-even-it/inequality-indonesia-millions-kept-poverty.

Padawangi, R., and M. Douglass. 2015. "Water, Water Everywhere: Toward Participatory Solutions to Chronic Urban Flooding in Jakarta". *Pacific Affairs* 88 (3): 517–550.

Pal, J. S., and E. A. Eltahir. 2016. "Future Temperature in Southwest Asia Projected to Exceed a Threshold for Human Adaptability". *Nature Climate Change* 6: 197–200.

Park, S. E., N. A. Marshall, E. Jakku, A. M. Dowd, S. M. Howden, E. Mendham, and A. Fleming. 2012. "Informing Adaptation Responses to Climate Change through Theories of Transformation". *Global Environmental Change* 22 (1): 115–126.

Parry, M., M. L. Parry, O. Canziani, J. Palutikof, P. van der Linden, and C. Hanson, eds. 2007. *Climate Change 2007: Impacts, Adaptation and Vulnerability*. Working group II contribution to the fourth assessment report of the IPCC. Vol. 4. Cambridge: Cambridge University Press.

Permanyer, I. 2013. "The Measurement of Success in Achieving the Millennium Development Goals". *The Journal of Economic Inequality* 11(3): 393–415.

Pelling, M. 2010. *Adaptation to Climate Change: From Resilience to Transformation*. Abingdon: Routledge.

Pido, E. J. 2017. *Migrant Returns: Manila, Development, and Transnational Connectivity*. Durham, NC: Duke University Press.

Philippines Statistics Authority. 2016. "Philippine Population Density (Based on the 2015 Census of Population)". *Philippines Statistics Authority*. Accessed 2 November 2018. https://psa.gov.ph/content/philippine-population-density-based-2015-census-population.

Prasad, N., F. Ranghieri, F. Shah, Z. Trohanis, E. Kessler, and R. Sinha. 2009. *Climate Resilient Cities: A Primer on Reducing Vulnerabilities to Disasters*. Washington, DC: World Bank. Accessed 26 November 2018. https://openknowledge.worldbank.org/handle/10986/11986.

Ragragio, J. 2003. *Urban Slums Report: The Case of Metro Manila Philippines*. Accessed 1 December 2018. www.ucl.ac.uk/dpu-projects/Global_Report/pdfs/Manila.pdf.

Sagala, S., J. Lassa, H. Yasaditama, and D. Hudalah. 2013. "The Evolution of Risk and Vulnerability in Greater Jakarta: Contesting Government Policy". *Institute for Resource Governance and Social Change*. Kupang: Institute for Resource Governance and Social Change.

Sahakian, M. D., and C. Dunand. 2014. "The Social and Solidarity Economy towards Greater Sustainability': Learning across Contexts and Cultures, from Geneva to Manila". *Community Development Journal* 50 (3): 403–417.

Sherwell, P. 2016. "$40bn to Save Jakarta: The Story of the Great Garuda". *Guardian*, 22 December. Accessed 2 November 2018. www.theguardian.com/cities/2016/nov/22/jakarta-great-garuda-seawall-sinking.

Thanvisitthpon, N., S. Shrestha, and I. Pal. 2018. "Urban Flooding and Climate Change: A Case Study of Bangkok, Thailand". *Environment and Urbanization ASIA* 9 (1): 86–100.

Transparency International. 2018. "Indonesia: An Overview of Corruption and Anti-Corruption". *Transparency International*. Accessed 26 November 2018. https://knowledgehub.transparency.org/helpdesk/indonesia-overview-of-corruption-and-anti-corruption.

United Nations. 2015. *The Millennium Development Goals Report.* United Nations. Accessed 26 November 2018. www.un.org/millenniumgoals/2015_MDG_Report/pdf/ MDG%202015%20rev%20(July%201).pdf.

United Nations. 2018. *Sustainable Development Goals.* Accessed 26 November 2018. www.un.org/sustainabledevelopment/sustainable-development-goals/.

UN Habitat. 2013. *State of the World's Cities 2012/2013: Prosperity of Cities.* Malta. Progress Press.

UN Habitat. 2016. *Slums Almanac 2015–16. Tracking Improvement in the Lives of Slum Dwellers.* Nairobi.

WCED (World Commission on Environment and Development). 1987. *Our Common Future.* Oxford: Oxford University Press.

Williams, G. A., B. Helmuth, B. D. Russell, Y. W. Dong, V. Thiyagarajan, V., and L. Seuront. 2016. "Meeting the Climate Change Challenge: Pressing Issues in Southern China and SE Asian Coastal Ecosystems". *Regional Studies in Marine Science* 8 (3): 373–381.

Win, T. L. 2017. "In Flood-Prone Jakarta, Will 'Giant Sea Wall' Plan Sink or Swim?" *Reuters,* 14 September. Accessed 26 November 2018. www.reuters.com/article/us-indonesia-infrastructure-floods/in-flood-prone-jakarta-will-giant-sea-wall-plan-sink-or-swim-idUSKCN1BP0JU.

Wise, R. M., I. Fazey, M. S. Smith, S. E. Park, H. C. Eakin, E. A. Van Garderen, and B. Campbell. 2014. "Reconceptualising Adaptation to Climate Change as Part of Pathways of Change and Response". *Global Environmental Change* 28: 325–336.

World Bank. 2012. *Thai Flood 2011: Rapid Assessment for Resilient Recovery and Reconstruction Planning: Overview.* Washington, DC: World Bank. Accessed 26 November 2018. http://documents.worldbank.org/curated/en/677841468335414861/Overview.

World Bank. 2013. *Using Participatory Mapping for Disaster Preparedness in Jakarta.* The World Bank. Accessed 26 November 2018. http://documents.worldbank.org/curated/ en/915261493797399464/pdf/114496-BRI-PILLAR-1-PUBLIC.pdf.

World Bank. 2015. *Indonesia – Second and Third Urban Poverty Project (English).* Washington, DC. Accessed 26 November 2018. http://documents.worldbank.org/curated/ en/491531467998527214/Indonesia-Second-and-Third-Urban-Poverty-Project.

World Bank. 2017. *Project Appraisal: Metro Manila BRT – Line 1 Project.* Accessed 2 November 2018. http://documents.worldbank.org/curated/en/270231488468381979/ pdf/Philippines-Metro-Manila-PAD-PAD1382-02272017.pdf.

World Population Review. 2018. *Jakarta Population 2018. World Population Review.* Accessed 2 November 2018. http://worldpopulationreview.com/world-cities/jakarta-population/.

Yarina, L. 2018. "Your Sea Wall Won't Save You: Negotiating Rhetorics and Imaginaries of Climate Resilience". *Places Journal,* March. Accessed 26 November 2018. https:// placesjournal.org/article/your-sea-wall-wont-save-you/?cn-reloaded=1.

Zhang, Y., D. Webster, A. Gulbrandson, A. G. Corpuz, A. Prothi, and J. C. Nebrija. 2014. *The Metro Manila Greenprint 2030: Building a Vision* (English). Washington, DC: World Bank Group. Accessed 26 November 2016. http://documents.worldbank.org/curated/ en/286861468189547797/The-Metro-Manila-greenprint-2030-building-a-vision.

8 Climate change, security, and sustainability

Simon Dalby

Sustainable development and environmental security

Thirty years after its publication, key themes from the World Commission on Environment and Development's (1987) report on *Our Common Future* have been formalized in the Sustainable Development Goals (SDGs), which are effectively the United Nations' policy blueprint for our times (United Nations 2015). But despite progress on many aspects of development, and the transformation of many aspects of the global economy in the last three decades, some of the key contradictions that "sustainable development" was attempting to ameliorate have been very persistent. One of the key premises in *Our Common Future*, which is often obscured in the focus on economy and environment, is the concern that environmental destruction will cause conflict, and in turn that conflict is a cause of environmental degradation. *Our Common Future* suggested that this would be especially disastrous in the event of a major nuclear war.

Sustainable development was portrayed as a necessity to avert conflict even if the precise modalities of how environmental change might cause conflict were never clearly specified in *Our Common Future*. The subsequent extensive discussion of environmental security since has been about both research into the relationships between resources, environmental change, and conflict, and about policy innovations that might facilitate peaceful co-operation and ameliorate economic shortages and difficulties that might in turn cause various forms of insecurity (Floyd and Matthews 2013). The concerns that local conflicts might escalate into wider-scale wars have persisted. In much recent American national security thinking, with the notable exception of those in the Trump administration, climate has become a priority concern in so far as it is seen as a stressor on societies that might be prone to conflict (CNA Corporation 2014). It has also become a concern to the US armed forces because many of their facilities, only most obviously the major naval base at Norfolk Virginia, are vulnerable to rising sea levels and increased storm intensity.

In so far as peripheral disruptions are viewed as a threat to metropolitan consumption – the implicit premise in much American thinking in particular – they obscure the larger causal relationships that point to both the direct disruptions of the global economy in terms of resource extraction, and the indirect disruptive effects of climate change (Dauvergne 2016). Getting this new context clear in discussions of environmental security is essential if the focus on sustainability is to shift from rural disruptions and poverty as a problem and focus instead on the cause of the disruptions in the global economy that are making so many people insecure in the face of rapid environmental change (Dalby 2014). The crucial point in rethinking environmental security in the twenty-first century is to recognize that current disruptions are caused by fossil-fuelled strategies of "development".

This all follows from the history of the growth of the global economy in the last few centuries, a matter of European imperial expansion initially and subsequently the "great acceleration" of a global industrial urban consumption economy (McNeill and Engelke 2016). The arguments that this economy is necessary for development, which is the prerequisite for dealing with poverty and other threats to human security, persist, despite the huge global inequalities and the continued marginalization of large parts of humanity who have not benefited from the largesse controlled by urban elites. The promise that eventually the poor will benefit from growth has been the argument at the heart of modern economics, notwithstanding repeated critiques about how huge inequalities persist despite growth (Piketty 2014).

Now the accelerating disruptions of climate change have made it clear that this economic model is untenable in the long run, dependent as it still is on huge-scale fossil fuel consumption (Holden *et al.* 2018). Securing energy supplies in these terms is ensuring climate disaster (Nyman 2018). Sustainable Development Goal 13, which promises urgent action on climate change is thus crucial to the achievement of all the other goals. This shift in perspective, from the traditional focus in development thinking on impoverished global peripheries, to the cause of climate disruption being fossil-fuel-driven metropolitan consumption, is key to any long-term effort to shape the planetary system in ways that allow sustainability to be developed as a common human future. This change of viewpoint focusing on humanity as a planetary scale actor shaping the future configuration of the biosphere is encapsulated in the now widespread use of the term Anthropocene to designate this new geological epoch (Lewis and Maslin 2018).

Where most of the SDGs look to the Global South (although ostensibly being "one world" goals), climate change and Goal 13 in particular require some serious policy rethinking in the Global North. Development funding, if rebranded as adaptation assistance under green development funds and other initiatives yet to come under the Framework Convention and the Paris Agreement, may help with some adaptations if it is scaled up in coming years (United Nations Environment Program 2018). However, if climate change is to be taken seriously then its causes need attention. These causes are primarily in the developed world's consumption of fossil fuels, and the landscape changes caused by resource extraction and the spread of industrial scale agriculture in rural areas to feed global markets. This chapter first discusses climate change before raising some pertinent issues about the contradictions in security thinking and in policymaking. It concludes that, in the current circumstances of the Anthropocene, developing sustainability requires a drastic cut in the use of combustion in the affluent metropoles of the global economy.

Sustainable Development Goals 13 and 16

The official UN list of the goals (United Nations 2015) has an asterisk attached to Goal 13 saying that the SDGs process is "Acknowledging that the United Nations Framework Convention on Climate Change is the primary international, intergovernmental forum for negotiating the global response to climate change". That said, it does specify the need to take urgent action to deal with climate change and its impacts. The targets refer to "strengthening resilience and adaptive capacity to hazards and natural disasters in all countries", integrating climate measures into national policies, improving education, awareness, and early warning. Target 13.a explicitly refers to raising $100bn annually by 2020 to address developing country needs on mitigation, transparency, and to "fully operationalize the Green Climate Fund through its capitalization as soon as possible". Target 13.b mentions the need to focus on climate change management "in least developed countries and small Island developing States, including focusing on women, youth and local and marginalized communities". But this green development fund is puny in comparison with the scale of fossil fuel subsidies world-wide (Coady *et al.* 2017).

Given the urgency of dealing with climate change – and Goal 13 is the only one that specifies its agenda as "urgent" – fairly drastic rethinking is clearly in order. The urgency comes not least because mitigation of climate change is mostly about preventing the burning of fossil fuels in the short-run because of their long-term consequences. The planetary system is already changing, as is especially evident in the Arctic as winter weather patterns are disrupted and ice cover dramatically reduced. Elsewhere rising sea levels, more severe weather, disasters, and agricultural disruptions are already happening as a result of climate change, hence the need to move quickly on climate change action. Rapidly cutting the use of fossil fuels should slow the process of change and facilitate adaptation (Steffen *et al.* 2018). But the slow progress under the United Nations Framework Convention on Climate Change (UNFCCC), despite the agreement in Paris in 2015 that action was needed soon to, in the language of Article 2 of the UNFCCC, "prevent dangerous anthropogenic interference with the climate system" (United Nations 1992, 4), presents major difficulties for the accomplishment of the other goals. It does so because climate change, with increasing unpredictable extreme weather events, is already an important disruptor of economic activities and a growing hazard to personal and infrastructural safety in many places.

Goal 16, on promoting peaceful and inclusive societies, focuses on promoting law and reducing all forms of violence, but does not explicitly make any links between environmental change and warfare, nor does it refer to the possibilities of drastic climate shocks causing international conflict. Despite many alarming headlines and media commentary on coming crises (Wallace-Wells 2017), most of the scholarly literature suggests that environmental change is unlikely to cause major conflicts in the near-term at least. There are some situations, related to agriculture and rivers supplying irrigation water as well as fish supplies, where adaptations in some places are likely to have aggravating consequences downstream in the absence of carefully worked out arrangements for "climate proofing" such things as water use agreements (Cooley and Gleick 2011).

Sustainable development relates directly to these matters of environmental security which, in the new circumstances of the Anthropocene, are going to need innovations in many places in the global system, not just among state governments. "The global commons cannot be managed from any national centre: The nation state is insufficient to deal with threats to shared ecosystems. Threats to environmental security can only be dealt with by joint management and multilateral procedures and mechanisms" (World Commission on Environment and Development 1987). Shifting the focus to agents other than national states suggests modes of political innovation that may be very helpful (Bernstein and Hoffman 2018); polycentric systems of governance are likely to be partly efficacious given the sheer diversity of places in social systems where decisions about combustion are made (Morrison *et al.* 2017). However, innovative solutions are needed to other social problems that are likely to impede the adoption of the innovative energy and land use systems required to tackle climate change.

Violent conflict remains a problem in many parts of the world, only most notably in the Middle East and Central Africa. These conflicts prevent many development issues from being dealt with effectively. SDG 16 attempts to tackle some of these issues, but not the matter of warfare, preparations for military action, nor the consequences of combat. War is directly destructive on battlefields, where it also frequently leaves a legacy of landmines and toxic wastes. It is indirectly a problem where it extracts resources in emergency mobilizations that ignore long-term consequences, and in such circumstances, as well as in routine preparations for combat, it diverts resources from providing social goods and services that are key to human security. Disease and deteriorating environmental quality are only some of the themes that violent conflict exacerbates. *Our Common Future* explicitly pointed to military expenditures as money that could be much better spent on other things. But while the 1990s did produce some "peace dividends", as the *Our Common Future* authors had hoped, spending on weapons and military actions by many states has crept upwards after the events of 9/11 and the remilitarization of international affairs.

Modern militaries also use large amounts of fossil fuels; in many states, they are the largest institutional user of fuels, and, as such, they directly exacerbate greenhouse gas emissions while their contributions are exempted from national inventories of greenhouse gas emissions. Rethinking energy and substituting clean electricity for combustion in numerous uses is now key to sustaining development in the global system. Numerous new technologies are emerging and changing energy systems (Patterson 2015). Many of these technologies offer very considerable promise of, to borrow an old phrase from the 1950s, "living better electrically". But these innovations are anathema to the fossil fuel industry and its vociferous advocates in Washington and elsewhere.

Firepower and national security

Related to this, although frequently less focused upon in the discussion, is the central contradiction at the heart of climate security issues: the simple fact that contemporary military actions are frequently about protecting the political order and the global economy built on burning the fossil fuels that are the primary cause of climate change. Using such "firepower" to protect economic and political power based on fossil fuels, and assuming that nature is a given context to be struggled over, rather than a context shaped by human economies, have been key to the geopolitical strategies of imperial powers in the last couple of centuries (Dalby 2018a). But as earth system science makes abundantly clear, assumptions of economic growth based on the ever-larger use of fossil fuels are untenable if a relatively stable biosphere is to be humanity's habitat in coming decades.

It is not at all obvious that the climate change negotiation process that culminated in Paris in 2015 has established a framework that can deliver either the goal of limiting average global temperature increase to "well below 2 degrees" (Celsius) above the preindustrial average (Falkner 2016), much less the aspirational target of 1.5 degrees necessary, so the models suggest, to keep sea level increase to a level that gives low-lying island countries some chance to survive. That being the case, viewed from the Pentagon, it is necessary to think through the security challenges that state inundation, and the rising severity and/or frequency of extreme events may have for the US military.

Viewed from those island states facing inundation from rising ocean levels things look very different. While they might welcome US assistance to deal with immediate consequences of storms, island state politicians cannot help but note the irony of the US military as one of the world's largest fossil-fuel users being the agency that deals with disasters. Hence, frequently leaders from the small island states insist that matters are framed as sustainable development, not as security (Dalby 2016a). The US military is also precisely the institution that has protected and indeed promoted the global economic system powered by fossil fuels that is causing climate change. Thus, there are contradictions at the heart of the system that links climate to security. While invoking the political language of climate action may suggest urgency, the institution closest to matters of emergency response is a key part of the problem.

The US military plans for the long-term and often frames things in terms of "shaping the future". These ideas of shaping probable future conflict spaces so that, if conflict occurs, it will happen in ways favourable to the US military, might by analogy be a useful formulation for thinking about climate. Both military and climate planning have to be thought about in terms of their long-term implications for future events. American military thinking, especially in terms of the long war against terrorism (Morrissey 2017), suggests that legal and political arrangements can shape the context for potential conflict, and, as such, the analogy with shaping policy frameworks to head off the worst impacts of climate change fits fairly closely.

Nonetheless, at least so far, climate has not become the priority "macro-securitization" in global politics – despite former President Obama's explicit invocation of climate as the most important security matter demanding global attention, laid out in his Brandenburg Gate speech in Berlin in June 2013. In Obama's (2013) words:

> With a global middle class consuming more energy every day, this must now be an effort of all nations, not just some. For the grim alternative affects all nations – more severe storms, more famine and floods, new waves of refugees, coastlines that vanish, oceans that rise. This is the future we must avert. This is the global threat of our time.

While this speech, presented where President Kennedy had made his key Cold War speech in the aftermath of the building of the Berlin Wall, led to heightened diplomatic activity, in particular with China, the climate has yet to become the taken-for-granted security priority.

As the Trump Administration's December 2017 *National Security Strategy* statement made clear, fears about other states, terrorism, and weapons of mass destruction, especially in the hands of potential adversaries in Tehran or Pyongyang, remain the priority despite clear indications that the climate is changing rapidly. While earth system scientists understand this as a potential phase shift in the earth system (Steffen *et al.* 2018), such thinking has yet to engage the imagination of many key financial or geopolitical thinkers in Washington and elsewhere. They still understand the world in terms of great power rivalries in a stable geographical context (Dalby 2018b). Or, if they do concede that major upheavals are afoot, these thinkers frequently fall back on assumptions that geoengineering and related technical fixes can be relied on to maintain the social order that has rendered them rich and powerful, and that has simultaneously formulated those who would challenge that social order as a problem of security. Those challenges are seen as external, threats from the Global South where climate is a threat multiplier or a catalyst of conflict requiring interventions to provide forms of environmental security that pre-empt violence and potential wars (Hardt 2018).

Conflict, environment and development

The subsidiary causes of climate change, in terms of rapid land use changes, only most obviously deforestation, the use of nitrogen-based fertilizers, and agricultural production systems (See Blay-Palmer and Young, Chapter 2) that generate methane too, are part and parcel of the transformation of the global biosphere that involves the widespread destruction of habitat, and the extermination of numerous species in what is a global-scale extinction event (Ceballos, Ehrlich, and Dirzo 2017). In the terms of the earth system science discussion, humanity is rapidly changing the geological conditions of the planet that have been so favourable to human flourishing in the last few thousand years (Gaffney and Steffen 2017). While humanity has long changed parts of its habitat and hunted other species to extinction, the sheer scale of current activities is so great that earth system scientists are now designating the present period as a new geological epoch, the Anthropocene (Davies 2016). This is the context in which sustainability now has to be rethought; old ideas of a given environmental context for development are no longer appropriate for the global disruptions underway. Discussions of who secures what, for whom, and where, have to be reformulated for a world where the assumptions of the use of ever more firepower to grow the economy and protect this system from potential disruptions can no longer be the basis for development or sustainability or security.

Many of the states and peoples who are most vulnerable to climate change are in the Global South, and getting climate adaptation to work in places where vulnerable people live is an unavoidable part of implementing the Paris Agreement and specifically the multiple national adaptation strategies involved. Viewing climate adaptation as a matter of engineering efforts and centralized impositions on vulnerable landscapes has already led to difficulties for both people and other species in affected areas (Sovacool and Linnér 2016). Here too the SDGs present a challenge to conventional understandings of development. While some development projects may facilitate adaptation, many of them are in danger of perpetuating policies that do not help the poor and marginal who, as climate change accelerates, are frequently directly in harm's way.

Rural conflict is not new. The expansion of commercial arrangements into subsistence systems has a long history of social disruption, not just in the obvious cases of European colonization and subsequent processes of modernization which have extended fossil-fuelled extraction systems into remote regions during the period of the great acceleration (Taylor 2015). Rob Nixon (2011) has termed the imposition of development and the resistance by local peoples to the conversion of traditional systems of rural economy into commercial ones a matter of the "environmentalism of the poor". Movements dedicated to trying to protect livelihoods in the face of the intrusion of external modes of property, appropriations, and enclosures relations face the "slow violence" of dispossession from traditional territories, pollution from mines and pesticides, flooding and water quality problems from dam building and irrigation schemes, as well as the destruction of traditional subsistence food sources, deforestation, and ecosystem fragmentation. All these processes are part and parcel of development as the expanding economic frontiers absorb ever more landscapes into the global economy while expelling local populations (Sassen 2014). In part this happens because Northern states and corporations are buying up land in Southern states to diversify sources of food as a mode of climate adaptation. This has disruptive consequences for rural property markets across the Global South (Dunlap and Fairhead 2014). These considerations of rural political economy are key to understanding the context in which climate adaptation plays out and are an overlooked factor in many formulations of climate migration and arguments that climate causes conflict.

Likewise, people displaced by droughts and weather disruptions are frequently impoverished in the process precisely because their wealth is tied into land holding, and when crops do not grow, income is lacking. Migration to cities in search of work is both an economic issue of impoverishment and, if land cannot be sold, a matter of an inability to access land elsewhere where it might be productive (Parenti 2011). In the much-cited case of Syria, climate-caused migration to cities is often portrayed as a cause of protests that lead to the civil war. But subsequent analyses have shown that climate change-induced migration and protests on the part of the dispossessed alone did not cause the civil war (Selby *et al.* 2017). There were numerous other stresses on Syrian society, not least the influx of refugees from Iraq and drought conditions that were aggravated by government policies removing fuel subsidies. These transformations of the rural political economy were a cause of much insecurity, if not a direct cause of the subsequent war.

In many places local populations resist the appropriation of land and resources and the deleterious actions of "development", whether it be the commercial appropriation of agricultural land, or the building of mines, dams, and pipelines. In the process they are frequently the victims of violence. The United Nations has a program on human rights and the environment which attempts to advocate for environmental rights and offer some protection for activists protesting against expulsion, environmental degradation, and displacement.[1] The *Guardian* newspaper in Britain has an ongoing project on "The Defenders" which highlights the work of human rights organizations tracking the killing of environmental defenders (Watts and Vidal 2017). Defending environments from economic encroachment is a very dangerous activity in many places, especially so when national governments blame international organizations and invoke claims of foreign illegal meddling to justify the repression of local activists (Matejova, Parker, and Dauvergne 2018). The death toll is clearly rising. Slow violence is now complemented by very fast violence among those who try to block environmentally damaging economic projects, a practice which Naomi Klein (2014) has termed "blockadia" given the frequency of the practice in diverse places, North and South.

The high profile "blockadia" protests at Standing Rock in North Dakota in 2017 focused attention on the attempts by indigenous peoples to invoke rights to land and territory as a mode of preventing fossil fuel extraction in the Global North. Noteworthy among the protestors were former United States veterans of the wars in the Middle East who came to assist the native protesters and their environmental movement allies. They did so to highlight the need to fundamentally change the energy system dependent on extracting fossil fuels and building potentially dangerous pipelines in numerous places, regardless of both the local impacts and the climate change consequences caused indirectly by the combustion of the hydrocarbons that flow though pipelines.

Climate adaptation and maldevelopment

Sovacool and Linnér (2016) itemise the dangers of inappropriate climate policies in terms of four key processes. Specifically, they argue that development frequently involves enclosures where public assets are transferred to private actors. Political acts of "exclusion" marginalize stakeholders and limit access to decision-making. Ecological "encroachment" frequently intrudes on ecosystems that are rich in biodiversity and that provide ecosystem services to larger environments. Finally, Sovacool and Linnér suggest that these difficulties are compounded by the entrenchment of social inequalities that further marginalize minorities and women. It should not be forgotten that these practices are frequently extensions of conventional development policies that are all about environmental change.

The crux of the issue is that sustainable development is supposed to shape environmental change in ways that are not deleterious to either human or non-human inhabitants of particular places, although the non-human life forms frequently get short shrift in these considerations. Thus, Sovacool and Linnér (2016) argue that, in so far as climate adaptation is treated as a traditional matter of development, such policies may make everything worse. The alternatives, ecologically focused strategies that take local context and social conditions seriously, rather than viewing rural areas from a viewpoint of metropolitan management for resource extraction or as a place for revenue generating carbon sinks, require thinking about economics very differently.

In so far as resistance to the expansion of modernity in rural areas links up with rural insurgencies and is then portrayed as a conflict multiplier or a catalyst for conflict, as in many North American formulations of climate security (CNA Corporation 2014), then there is the danger that policy actions designed to shore up political stability, whether justified in terms of climate adaptation or not, will have counter-productive consequences. Research work in the 1990s has clearly shown that conflict related to environmental matters is frequently related to misplaced modes of development or "maldevelopment" (Baechler 1998). If such policies are funded by Northern states as part of climate adaptation measures through various strategies of ill-considered green development, the dangers are that they will cause resistance and "backdraft effects" that are counter-productive to both local ecological flourishing and international policy development (Dabelko *et al.* 2013).

Many of these are the result of failures to think through the local ecological contexts within which adaptation projects are undertaken, a matter of centralized development priorities and economic modes that displace local peoples or, as in the case of climate adaptation-induced migration in Vietnam, build such things as dykes to prevent flooding which also do not allow necessary silt to accumulate on fields (Chapman and Tri 2018). In Bangladesh, government efforts to link development and climate adaptation often end up dispossessing the poorest people who lack formal legal title to marginal lands in danger of inundation (Sovacool and Linnér 2016). Conflicts caused by such modes of

development are simply aggravating the difficulties rather than facilitating preparation to deal with increasingly severe disruptions to ecology and economy
(Risi 2017).

Viewed in these terms, SDGs 13 on Climate and 16 on Justice and Institution Building suggest the need to think through responses to climate change in
more dramatic ways than much of the conventional discussion has so far considered. Building resilience into social and economic systems to deal with more
extreme weather disruptions and the immediate physical dangers, as well as the
economic dislocations that result, is part of what needs to be done. Unfortunately, however, resilience thinking frequently does not offer visions of more
robust futures or deal with how to "build back better" very effectively (Grove
2018). Mark Pelling's (2011) formulation suggests that it is now necessary to
think about all this in terms of transformation, remaking societies to simultaneously prepare for the future in a warmer world while making improvements to
existing economic and social conditions. Taking the SDGs together as a whole
package, rather than focusing on them individually, seems to suggest precisely
this, and the SDGs program adopts the terminology of transformation to highlight the need for much more than incremental change.

Doing all this while recognising the interconnectedness of the global economy
and the inequities in the system is obviously a very tall order; much more than
traditional ideas of conservation and resource management are needed. Societies
are frequently struggling with immediate priorities that make long-term planning
a luxury, even where notions of justice are part of the political discussion of ecopolitics (Stoett 2012). Nonetheless, thinking through ecological limits and the
possibilities for radical change are now unavoidable, as the imperatives of earth
system analyses point to the limits of fossil-fuelled economic expansion if sustainability is to be taken seriously (Holden *et al.* 2018). The contradictions at the
heart of all this also point to the necessity of rethinking security and delinking it
from the perpetuation of a global system that extracts resources from rural areas,
generates vast quantities of waste, and equates security with the preservation of a
system that is obviously unsustainable.

Securing the future

Climate science, and earth system thinking more generally, is increasingly clear
in terms of the potential for disasters of many sorts if the current trajectory of
greenhouse gas emissions is not changed, and soon. Numerous reports have suggested that we live in turnaround decades or face imminent tipping points
beyond which the earth system will enter some new configuration, one much
less likely to be conducive to human civilization (Steffen *et al.* 2018). Invoking
disaster, however, frequently leads to a sense of hopelessness or an emergency
mobilization of resources to combat a danger understood as external to whatever
entity is threatened. But the ongoing struggles in rural areas concerning the
forms of development that are being perpetuated by the global economy suggest
a very different set of priorities for climate action and the implementation of

the SDGs. Ideas of post-development, "degrowth", and transformation are powerfully reinforced when they are connected up with the earth system discussion and the need to think about new modes of economy after fossil fuels, i.e. ones that simultaneously attend to human social needs without violating earth system boundaries (Raworth 2017).

The earth system analysis and the formulation in terms of the Anthropocene, as a new era being made by humanity, present some promising possibilities to challenge the doom and disaster framings of the future (Dalby 2016b). But if it is going to be an effective form of imagination, it will have to focus on matters of production and economy rather than on imminent disaster. The decisions that matter are ones that focus on the long-term consequences of investments, on making solar panels and wind turbines rather than carbon dioxide, on social innovations and on modes of economy that do not assume ever larger material throughputs and "growth" as the solution to all social ills.

The divestment movement in particular suggests that, while war and struggle are part of the political rhetoric involved in imagining a sustainable future (Mangat, Dalby, and Paterson 2018), the key "struggles" are to direct capital, and crucially public policy, toward shaping economic decisions that make things that do not disrupt the climate. This extends also to the critical land-use decisions that affect biodiversity (Ceballos, Ehrlich, and Dirzo 2017). Whether we make landscapes for monoculture agriculture or for biodiverse, polyculture production systems have important repercussions for future generations. Understanding ecology and the global economy as a matter of deciding what kinds of landscapes will be made, either extractivist ones solely to feed a global economy with commodities demanded by short-term pricing priorities (Buxton and Hayes 2016), or ones that take seriously biodiversity protection, agricultural flexibility as well as the capacity of complex ecosystems to more effectively ameliorate climate extremes, now matters greatly.

Agreements among leading corporations and government arrangements to rapidly phase out the most deleterious production systems will help. Pricing fuels appropriately by using carbon taxes and removing petroleum subsidies are policies designed to move economies away from greenhouse gas emissions. The larger focus has to be on production, and on making things that do not involve combustion on the one hand and that facilitate the reduction of carbon dioxide in the atmosphere on the other. The overall objective is ending the fossil fuel system (Princen, Manno, and Martin 2015). This is not likely to work in terms of centralized authoritarian politics on the global scale, although it will certainly help if some universal environmental regulations such as agreements to stop using coal-powered electricity generating systems, are crafted and enforced to restrict, if not simply ban, the use of dangerous substances. This has been done in the case of CFCs and ozone, and it makes sense to consider such innovations as designating coal in particular a restricted or controlled substance both because of its pollution and its climate change effects (Burke *et al.* 2016). Developing such things as international norms that render combustion in many forms an inappropriate activity is now a priority.

Climate security: combining Goals 13 and 16

But for such norms to be effective the hierarchical geopolitical arrangements, those epitomized by the violent geopolitical use of firepower in the twentieth century, will have to be constrained by novel institutions, both in terms of international regimes and local innovations to make a post carbon-fuelled economy that facilitates ecological diversity (Biermann 2014). This requires thinking in imaginative ways and explicitly foregrounding "the planetary" in political discourse, rather than assuming that development models based on combustion can be the basis of future policies.

A number of institutions have been responding to the challenge of climate security and, in so far as they influence policy in coming years, they will move the SDGs agenda forward, helpfully. In doing so these initiatives require a refocusing of security away from traditional ideas of violent responses to supposedly external threats and towards an active caring for both people and ecologies as constituent parts of an evolving system (Harrington and Shearing 2017). This requires taking the impetus behind *Our Common Future* and thinking in terms of peacebuilding and constructive engagement with rural communities facing climate change and economic disruption simultaneously. While this is a very ambitious agenda, both warfare and climate change are serious obstacles to sustainable development and tackling them together makes good policy sense.

One major report instigated by the G7 in 2015 bluntly posed the issue as the need for "A New Climate for Peace" (G7 2015). This suggested clearly that climate risks were urgent given the fragility of many states and their vulnerabilities to climate disruptions in coming decades. Building on this, more recently "The Planetary Security Initiative",[2] a consortium of leading think tanks and researchers launched by the Dutch government contemporaneously with the SDGs, exemplifies the attempt to link climate adaptation with peacebuilding. Thus, the agenda laid out in these initiatives aims to tackle the issues of conflict prevention as part of the sustainable development agenda. As *Our Common Future* long ago suggested, to be effective these have to be done in conjunction.

What is clear too is that there is a need to link up these initiatives with the new energy systems that are becoming available. The new micro-grids and possibilities for bypassing large-scale centralized twentieth century generating systems are changing the landscape for rural development and ushering in numerous possibilities for buildings that do not rely on extensive external energy inputs to function. But to make these the basis for a new economy that allows humanity to flourish within a relatively stable global biosphere will require a focus on investing in these novel systems quickly, while phasing out coal-powered electricity generation and petroleum-based transport. The earth system analyses of the Anthropocene make it clear that nothing less than the elimination of fossil fuels in coming decades will be enough if the promise in *Our Common Future* is to be realized.

Acknowledgements

Funding to support this research comes from the Canadian Social Sciences and Humanities Research Council grant on "Borders in Globalization" (895-2012-1020).

Notes

1 See the UN Special Rapporteur on Human Rights and the Environment's website: http://srenvironment.org/. Accessed 24 October 2018.
2 For more information on this initiative, see www.planetarysecurityinitiative.org/. Accessed 24 October 2018.

References

Baechler, G. 1998. "Why Environmental Transformation Causes Violence: A Synthesis". *Environmental Change and Security Project Report* 4: 24–44.

Bernstein, S., and M. Hoffman. 2018. "The Politics of Decarbonization and the Catalytic Impact of Subnational Climate Experiments". *Policy Sciences* 51 (2): 189–211.

Biermann, F. 2014. *Earth System Governance*. Cambridge: MIT Press.

Burke, A., S. Fishel, A. Mitchell, S. Dalby, and D. Levine. 2016. "Planet Politics: A Manifesto from the End of IR". *Millennium* 44 (3): 499–523.

Buxton, N., and B. Hayes, eds. 2016. *The Secure and the Dispossessed: How the Military and Corporations are Shaping a Climate-Changed World*. London: Pluto.

Ceballos, G., P. R. Ehrlich, and R. Dirzo. 2017. "Biological Annihilation via the Ongoing Sixth Mass Extinction Signalled by Vertebrate Population Losses and Declines". *Proceedings of the National Academy of Sciences* 114 (30): E6089–6096.

Chapman, A., and V. P. D. Tri. 2018. "Climate Change is Triggering a Migrant Crisis in Vietnam". *The Conversation*, 9 January. Accessed 25 October 2018. https://theconversation.com/climate-change-is-triggering-a-migrant-crisis-in-vietnam-88791.

CNA Corporation. 2014. *National Security and the Accelerating Risks of Climate Change*. Alexandria: CNA Corporation.

Coady, D., I. Parry, L. Sears, and B. Shang. 2017. "How Large Are Global Fossil Fuel Subsidies?". *World Development* 91: 11–27.

Cooley, H., and P. H. Gleick. 2011. "Climate Proofing Transboundary Water Agreements". *Hydrological Sciences Journal* 56 (4): 711–718.

Dabelko, G., L. Herzer, S. Null, M. Parker, and R. Sticklor. 2013. "Backdraft: The Conflict Potential of Climate Change Adaptation and Mitigation". *Woodrow Wilson Center Environmental Change and Security Program Report* 14 (2): 1–60.

Dalby, S. 2009. *Security and Environmental Change* Cambridge: Polity.

Dalby, S. 2014. "Environmental Geopolitics in the Twenty First Century". *Alternatives: Global, Local, Political* 39 (1): 3–16.

Dalby, S. 2016a. "Climate Change and the Insecurity Frame". In *Reframing Climate Change: Constructing Ecological Geopolitics*, edited by S. O'Lear and S. Dalby, 83–99. London: Routledge.

Dalby, S. 2016b. "Framing the Anthropocene: The Good, the Bad, and the Ugly". *The Anthropocene Review* 3 (1): 33–51.

Dalby, S. 2018a. "Firepower: Geopolitical Cultures in the Anthropocene". *Geopolitics* 23 (3): 718–742.

Dalby, S. 2018b. "Geopolitics in the Anthropocene". In *The Return of Geopolitics*, edited by A. Bergeson and C. Suter, 149–166. Zurich: Lit.

Dauvergne, P. 2016. *Environmentalism of the Rich*. Cambridge: MIT Press.

Davies, J. 2016. *The Birth of the Anthropocene*. Berkeley: University of California Press.

Dunlap, A., and J. Fairhead. 2014. "The Militarisation and Marketisation of Nature: An Alternative Lens to 'Climate-Conflict'". *Geopolitics* 19 (4): 937–961.

Falkner, R. 2016. "The Paris Agreement and the New Logic of International Climate Politics". *International Affairs* 92 (5): 1107–1125.

Floyd, R., and R. Matthew, eds. 2013. *Environmental Security: Approaches and Issues*. London: Routledge.

Gaffney, O., and W. Steffen. 2017. "The Anthropocene Equation". *The Anthropocene Review* 4 (1): 53–61.

Grove, K. 2018. *Resilience*. New York: Routledge.

G7. 2015. "A New Climate for Peace: Taking Action on Climate and Fragility Risks". Executive Summary. Accessed 24 October 2018. www.newclimateforpeace.org.

Hardt, J. 2018. *Environmental Security in the Anthropocene: Assessing Theory and Practice*. London: Routledge.

Harrington, C., and C. Shearing. 2017. *Security in the Anthropocene: Reflections on Safety and Care*. Bielefeld: Transcript.

Holden, E., K. Linnerud, D. Banister, V. J. Schwanitz, and A. Wierling. 2018. *The Imperatives of Sustainable Development: Needs, Justice, Limits*. London: Routledge/Earthscan.

Klein, N. 2014. *This Changes Everything: Capitalism vs. The Climate*. Toronto: Knopf.

Lewis, S., and M. Maslin. 2018. *The Human Planet: How We Created the Anthropocene*. London: Pelican Books.

Mangat, R., S. Dalby, and M. Paterson. 2018. "Divestment Discourse: War, Justice, Morality and Money". *Environmental Politics* 27 (2): 187–208.

Matejova, M., S. Parker, and P. Dauvergne. 2018. "The Politics of Repressing Environmentalists as Agents of Foreign Influence". *Australian Journal of International Affairs* 72 (2): 145–162.

McNeill, J. R., and P. Engelke. 2016. *The Great Acceleration: An Environmental History of the Anthropocene since 1945*. Cambridge: Harvard University Press.

Morrison, T. H., N. Adger, K. Brown, M. C. Lemos, D. Huitema, and T. P. Hughes. 2017. "Mitigation and Adaptation in Polycentric Systems: Sources of Power in the Pursuit of Collective Goals". *WIREs Climate Change* 8: 1–16.

Morrissey, J. 2017. *The Long War*. Athens: University of Georgia Press.

Nixon, R. 2011. *Slow Violence and the Environmentalism of the Poor*. Cambridge: Harvard University Press.

Nyman, J. 2018. *The Energy Security Paradox: Rethinking Energy (In)security in the United States and China*. Oxford: Oxford University Press.

Obama, B. 2013. "Remarks by President Obama at the Brandenburg – Berlin, Germany". *The White House*, 19 June. Accessed 25 October 2018. https://obamawhitehouse. archives.gov/the-press-office/2013/06/19/remarks-president-obama-brandenburg-gate-berlin-germany.

Parenti, C. 2011. *Tropic of Chaos: Climate Change and the New Geography of Violence*. New York: Nation Books.

Patterson, W. 2015. *Electricity vs. Fire: The Fight for our Future*. Amersham: Walt Patterson.

Pelling, M. 2011. *Adaptation to Climate Change: From Resilience to Transformation*. London: Routledge.

Piketty, T. 2014. *Capital in the Twenty First Century*. Cambridge: Harvard University Press.

Princen, T., J. P. Manno, and P. L. Martin, eds. 2015. *Ending the Fossil Fuel Era*. Cambridge: MIT Press.

Raworth, K. 2017. *Doughnut Economics: Seven Ways to Think like a 21st Century Economist*. London: Random House.

Risi, L. H. 2017. "Backdraft Revisited: The Conflict Potential of Climate Change Adaptation and Mitigation". *New Security Beat*, 12 January. Accessed 25 October 2018. www.newsecuritybeat.org/2017/01/backdraft-revisited-conflict-potential-climate-change-adaptation-mitigation/.

Sassen, S. 2014. *Expulsions: Brutality and Complexity in the Global Economy*. Cambridge: Harvard University Press.

Selby, J. O. S. Dahi, C. Frolich, and M. Hulme. 2017. "Climate Change and the Syrian Civil War Revisited". *Political Geography* 60 (2017): 232–244.

Sovacool, B. K., and B-O. Linnér. 2016. *The Political Economy of Climate Change Adaptation*. London: Palgrave Macmillan.

Steffen, W., J. Rockström, K. Richardson, T. M. Lenton, C. Folke, D. Liverman, C. P. Summerhayes *et al.* 2018. "Trajectories of the Earth System in the Anthropocene". *Proceedings of the National Academy of Sciences* 115 (33): 8252–8259.

Stoett, P. 2012. *Global Ecopolitics: Crisis, Justice and Governance*. Toronto: University of Toronto Press.

Taylor, M. 2015. *The Political Ecology of Climate Change Adaptation: Livelihoods, Agrarian Change and the Conflicts of Development*. London: Routledge/Earthscan.

United Nations. 1992. *United Nations Framework Convention on Climate Change*. FCCC/INFORMAL/84. Accessed 26 October 2019. https://unfccc.int/resource/docs/convkp/conveng.pdf.

United Nations. 2015. *Transforming Our World: The 2030 Agenda for Sustainable Development*. A/RES/70/1. Accessed 24 October 2018. https://sustainabledevelopment.un.org/post2015/transformingourworld/publication.

United Nations Environment Program. 2018. *Making Waves: Aligning the Financial System with Sustainable Development*. Accessed 25 October 2018. http://unepinquiry.org/making-waves/.

Wallace-Wells, D. 2017. "The Uninhabitable Earth". *New York Magazine*, 9 July.

Watts, J., and J. Vidal. 2017. "Environmental Defenders being Killed in Record Numbers". *Guardian*, 13 July.

World Commission on Environment and Development. 1987. *Our Common Future*. Oxford: Oxford University Press.

9 Development as a determinant of climate risk and policy challenge

Vanessa Schweizer

Introduction

Risk is conventionally defined as the product of impact and probability. When it comes to climate change, impacts can manifest in many ways such as the loss of lives, species, or livelihoods, properties damaged, or some combination of these (IPCC 2014). Often, people think of climate risk deterministically, where the amount of greenhouse gases in the atmosphere determines the level of climate change. However, at current levels of emissions and the observed change in global average temperature, this is only half the story.[1]

The Special Report "Managing the Risks of Extreme Events and Disasters to Advance Climate Change Adaptation" (IPCC 2012) published by the Intergovernmental Panel on Climate Change, or IPCC, usefully visualizes climate risk as the *intersection* between changes in physical climate and socio-economic conditions experienced by humans. It refers to how economic development policy shapes the socio-economic realities of people potentially in harm's way (e.g. those exposed to hurricanes or vulnerable to flooding due to sea-level rise). Thus, climate risk is not entirely preordained by the state of Earth's atmosphere. As is the case with risky activities that we undertake each day (e.g. walking across a busy street), *choices* that people make to act pre-emptively versus obliviously (e.g. deciding to pay attention to oncoming traffic rather than to one's cell phone) will also determine risk.

Since the publication of the aforementioned Special Report, the scientific community developed a new, two-part analytical framework to make the effects of policy choices on climate risk more apparent. Historically, climate impact studies on sectors of interest such as agricultural productivity, water availability, or public health tended to consider only the hypothetical effects of climate change on areas or populations similar to the present day (Birkmann *et al.* 2013). This likely constrained the comprehensiveness of the research, since socio-economic characteristics such as the size of populations, their spatial distribution, their ages, incomes, etc. are unlikely to remain constant over the next 30, 50, or 80 years. To correct this widespread misconception, the latest generation of scenarios purposely separates climate scenarios (which are based on emissions trajectories, or Representative Concentration Pathways; see van

Vuuren *et al.* [2011]) from socio-economic scenarios (which are based on Shared Socio-economic Pathways; see Nakicenovic, Lempert, and Janetos [2014]). For risk analysis, such a scenario design makes it possible to hold a particular assumption about the amount of climate change constant but systematically vary socio-economic conditions (van Vuuren *et al.* 2014). This makes it possible to distinguish what amount of climate risk is attributable to climatic versus non-climatic factors.[2] This chapter briefly explains two insights emerging from the two-part framework and points out Sustainable Development Goals (SDGs) embedded within the Shared Socio-economic Pathways (SSPs). The chapter concludes with a discussion of SDGs that should be prioritized based on the climatic and economic impacts of failing to achieve particular SDGs in alternative SSPs.

Insight 1: socio-economic development matters for both emissions and climate adaptation

SDGs are examples of non-climatic factors that matter for reaching climate goals in the UN Framework Convention on Climate Change. Conceivably, future economic development paths could be wildly successful with achieving all SDGs, which would be ideal. Alternatively, they might prioritize increasing per capita income in the short run without concern for the climate risks associated with continuing to use and extract fossil fuels. Inequality is also important: underinvesting in human development, such as through public education, would be associated with higher population growth in poorer countries, which ultimately has negative impacts on achieving the SDGs and can put larger populations in harm's way under a changing climate. This means that alternative priorities for how economies pursue development and growth will have implications for both what level of climate change occurs and how well societies adapt to it.

To systematically explore such implications, the SSPs describe five alternative plausible scenarios that are distinguished by their respective mixes of socio-economic traits (Figure 9.1 and Table 9.1). *Ceteris paribus*, these traits are expected to make it easier or harder to mitigate emissions or to adapt to climate change. SSPs are defined both qualitatively (O'Neill *et al.* 2017) and with quantitative projections, which are available from a database (IIASA 2016). The details of the SSPs were informed by literature review, integrated assessment modelling (van Vuuren *et al.* 2012), and expert judgment (e.g. O'Neill *et al.* 2012; Schweizer and O'Neill 2014). For each SSP, there are contrasts between whether various SDGs are achieved, how quickly, and by what means.

SSPs purposely span a range of alternatives as to how the future could play out in the absence of explicit climate policy. Across their qualitative descriptions and narratives, all 17 SDGs are reflected as shown in Table 9.1.[3] Figure 9.1 provides a visual summary comparing progress on various SDGs across the five SSPs. For further exemplification, highlights of SSP narratives are in Table 9.1. The qualitative descriptions in Table 9.1 are from O'Neill *et al.* (2017). From

Table 9.1 Abridged qualitative descriptions of Shared Socio-economic Pathways

	SSP1: taking the green road (sustainability)	SSP2: middle of the road	SSP3: a rocky road (regional rivalry)	SSP4: a road divided (inequality)	SSP5: taking the highway (fossil-fuelled development)
Demographics					
Population growth	Relatively low	Medium	Low in OECD countries, high elsewhere	Low in OECD countries, high elsewhere	Relatively low
Mortality [SDGs 3, 2]	Low	Medium	High	High in high-fertility countries, medium elsewhere	Low
Urbanization level	High	Medium	Low	Medium in rich OECD countries, high elsewhere	High
Urbanization type [SDG 11]	Well managed	Historical pattern continues	Poorly managed	Mixed across, within cities	Better management over time, some sprawl
Human Development					
Education [SDG 4]	High	Medium	Low	Within regions, unequal	High
Gender equality [SDG 5]	High	Medium	Low	Unequal within regions medium trend at best	High
Equity [SDG 10]	High	Medium	Low	Medium	High
Access to health facilities, water, sanitation [SDGs 3, 6, 9]	High	Medium	Low	Unequal within regions; medium trend at best	High

Economy and Lifestyle					
Growth (per capita) [SDG 8]	High in developing countries, medium elsewhere	Medium, uneven	Slow	Low in developing countries, medium elsewhere	High
Inequality [SDGs 10, 1]	Reduced across, within countries	Uneven moderate reduction across, within countries	High, especially across countries	High, especially within countries	Strongly reduced, especially across countries
Globalization	Connected markets, regional production	Semi-open globalized economy	De-globalizing; focus on regional security	Globally connected elites	Strongly globalized, increasingly connected
Consumption and diet [SDG 12]	Decreasing materialism, low meat consumption	Material-intensive consumption, medium meat consumption	Material-intensive consumption	Elites: high consumption; Rest: low consumption, low mobility	Materialism, status consumption, tourism, high mobility, meat-rich diets
Policies and Institutions [SDGs 16, 17]					
International cooperation	Effective	Relatively weak	Weak, uneven	Effective for globally connected economy, not for vulnerable populations	Effective pursuit of development goals; more limited for environment
Dominant policy orientation/priority	Sustainable development	Weak focus on sustainability	Security	Benefits for the political and business elite	Toward development, free markets, human capital
Institutional effectiveness	Effective at international, national levels	Uneven, modest	Weak global institutions; national governments dominate	Effective for political and business elite, not for others	Increasingly effective; oriented toward fostering competitive markets

continued

Table 9.1 Continued

	SSP1: taking the green road (sustainability)	SSP2: middle of the road	SSP3: a rocky road (regional rivalry)	SSP4: a road divided (inequality)	SSP5: taking the highway (fossil-fuelled development)
Technology and Environment					
Tech development [SDG 9]	Rapid	Medium, uneven	Slow	Rapid in high-tech economies, sectors; slow in others	Rapid
Tech transfer [SDGs 7, 3, 9]	Rapid	Slow	Slow	Little transfer within countries to poor populations	Rapid
Tech change in the energy sector [SDGs 7]	Toward efficiency, renewables	Some investment in renewables, but reliance on fossil fuels	Slow, directed toward domestic energy resources	Diversified investments in efficiency, low-carbon supply	Toward fossil fuels; alternatives not pursued
Constraints on fossil fuels [SDG 13]	Preferences shift away from fossil fuels	No reluctance to use unconventional resources	Unconventional resources, focus on domestic supply	Anticipated constraints drive prices up; increase volatility	None
Environment [SDGs 14, 15][a]	Improving conditions over time	Continued degradation	Serious degradation	Highly managed, improved near more prosperous areas; degraded otherwise	Highly engineered, successful management of local issues

Source: Adapted from O'Neill et al. (2017a). Sustainable Development Goals (SDGs) have been added to scenario factors in the far left column.

Note

a SSPs could be interpreted to deduce progress for SDG14, conservation and sustainable use of the ocean and marine resources. However, the research discussed in this chapter focuses on impacts to terrestrial ecosystems, SDG15.

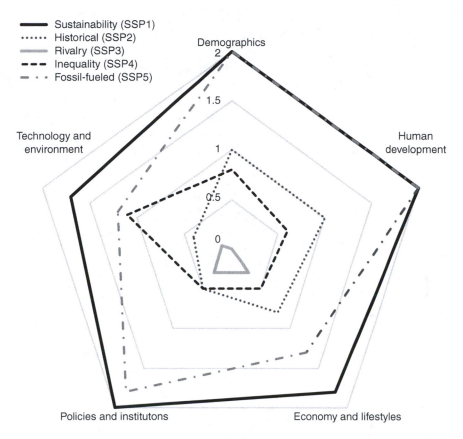

Figure 9.1 Progress on the Sustainable Development Goals across Shared Socio-economic Pathways.

Source: created by author. Scores in the plot are ordinal, where 2 = achievement, 1 = progress, 0 = stagnation. In a clockwise direction, axes correspond to the categories of scenario factors below in Table 9.1 (e.g. Demographics, Human Development as categorized by O'Neill *et al.*, 2017a. "Technology & Environment" is a further aggregation of categories). Decimal scores reflect partial or uneven progress on SDGs within a category.

Figure 9.1, progress on the SDGs clearly differs across SSPs as indicated by differences in the area that they cover. They also differ with respect to what SDGs are prioritized. Plots resembling regular polygons (e.g. "Sustainability") indicate that most SDGs are pursued equally. Plots that are lopsided in some way indicate that only one or a few dimensions are prioritized.

The "Historical Trends" scenario is referred to in the literature as SSP 2. Generally speaking, it describes a future where some progress is made on the SDGs (see the ordinal scores of 1 in Figure 9.1). However, similar to the Millennium Development Goals, the SDGs ultimately may or may not be achieved. "Historical Trends" reflects pessimistic outlooks on SDGs related to consumption (i.e. the Economy and Lifestyle dimension in Figure 9.1), and weak international co-operation and institutional effectiveness (i.e. the Policies and Institutions dimension in Figure 9.1). Slow technology transfer, reliance on fossil fuels, no reluctance to use unconventional fossil fuels such as tar sands, and continued environmental degradation slow progress on the dimension of Technology and Environment.

From the "Historical Trends" base case, the remaining SSPs describe alternative directions policymakers might take. "Sustainability" (a.k.a. SSP 1) describes excellent progress on all SDGs. Inequality is reduced across and within countries, but due to slower economic growth in comparison to "Fossil-fuelled Development",[4] inequality may not be reduced to the maximal extent (see the Economy and Lifestyle dimension in Figure 9.1). Great progress is also made in Technology and Environment; however, by 2030, work is likely to remain to shift the global energy system fully away from fossil fuels.

In contrast, "Fossil-fuelled Development" (SSP 5) describes a future where many SDGs are achieved through continued use of fossil fuels (in Table 9.1, see the categories Demographics and Human Development respectively for SDGs 1–11; see the category Policies and Institutions for SDGs 16–17). SDGs specifically related to the environment and climate are largely ignored (in Table 9.1, see the category Economy and Lifestyle for SDGs 12–13 related to consumption and emissions, i.e. preferences for materialism, meat consumption, and personal mobility), or they are supported in a highly managed fashion (in Table 9.1, see the category Technology and Environment for SDGs 14–15 related to ocean and terrestrial ecosystem health). In recognition of competing interpretations of what it means to implement sustainable development (Sanwal 2012), "Fossil-fuelled Development" raises an interesting question: To the chagrin of those concerned about planetary boundaries (Steffen *et al.* 2015), how feasible might it be to prioritize a strong push to improve human development (SDGs 1–11, 16–17) at the expense of the environment?

Compared to "Historical Trends", the SSP focusing on "Inequality" (SSP 4) depicts a mixed future, where some populations within and across countries are better off, but many remain behind. The energy sector sees vibrant change, and more prosperous communities and countries focus on improved environmental management near their own borders (progress on the Technology and Environment dimension in Figure 9.1). However, progress on SDGs in other dimensions is comparable to or worse than "Historical Trends". Mortality remains high in high-fertility countries (Demographics dimension in Figure 9.1). Improvements stall or reverse in gender equity as well as for access to education, health facilities, water, and sanitation for vulnerable populations (Human Development dimension in Figure 9.1). Economic growth stalls or reverses in developing countries, and inequality is high within countries, with consumption patterns resembling the Gilded Age (Economy and Lifestyle dimension in Figure 9.1). In short, "Inequality" describes a dynamic future for high-tech, high-return sectors and the global elites participating in them, while other classes are much more insecure. With its dominant policy orientation focused on benefits for elites, welfare states and progressive reforms may also be weakened (or, in developing countries, squelched before they become established).

"Regional Rivalry" (SSP 3) is characterized by de-globalization and a focus on regional or domestic security. Because of this inward focus, it reflects the least progress on the SDGs, with virtually no progress on some dimensions in Figure 9.1. Mortality remains high and urbanization is poorly managed (Demographics dimension). Investments in Human Development (e.g. education, access to health facilities) are low across the board. Inequality is high across countries, consumption is inefficient and material-intensive, and global economic growth trends are the slowest across the five SSPs (Economy and Lifestyle dimension). International policies and institutions are also weaker than "Historical Trends". Technology change in the energy sector focuses on domestic energy resources, corresponding to some improvement on SDG 7 (Technology and Environment dimension). However, for all other SDGs related to Technology and Environment, there are modest or virtually no improvements.

The five SSPs were developed to standardize socio-economic assumptions across various climate impact studies. However, they are sufficiently detailed that they can also be used to envision progress on the SDGs.

Insight 2: truly sustainable development must be climate resilient

Article 2 of the UN Framework Convention on Climate Change (UNFCCC) states, "The ultimate objective of this Convention … is to achieve … stabilization of greenhouse gas concentrations in the atmosphere that would prevent dangerous anthropogenic interference with the climate system" (United Nations 1992).

This speaks directly to the objective of decreasing climate risk. However, what the stabilization target for emissions should be depends on the amount of climate change to which human systems can adapt.[5] Drivers of "anthropogenic interference with the climate system" correspond with challenges for mitigation such as technological choices for energy and land use, political and institutional priorities for hastening (or not) energy transitions, resource efficiencies, and lifestyles. However, the socio-economic traits related to these drivers simultaneously affect challenges for adaption, namely exposure to potential climate hazards through the spatial distribution of people, property, and economic activities as well as their vulnerability. Naturally, decreasing challenges to *both* mitigation and adaptation decreases climate risk the most. "Climate-resilient pathways" refer to development trajectories that combine adaptation policy, mitigation policy, and effective institutions for sustainable development (Denton *et al.* 2014). A feature of the SSPs is that risk and policy analysts can compare the best-case scenario ("Sustainability", SSP 1) and the worst-case scenario ("Regional Rivalry", SSP 3) not only to the "Historical Trends" SSP 2 but also to examples of "second-best worlds" described by "Fossil-fuelled

Development" (SSP 5) and "Inequality" (SSP 4). In short, the SSPs provide clues for trade-offs incurred by focusing selectively on certain SDGs or on certain world regions rather than making strong progress on all SDGs for all countries.

In this section, the climate resilience for each SSP is discussed and compared. In a series of papers that appeared in *Global Environmental Change*, the SSPs were quantified through integrated assessment modelling (van Vuuren *et al.* 2017a). Figure 9.2 summarizes trade-offs across the SSPs for their climate change outcomes and estimated costs for emissions mitigation. For the purpose of comparing their respective climate resilience, "Sustainability" (SSP 1) is examined first, followed by the "second-best" worlds of "Fossil-fuelled Development" (SSP 5) and "Inequality" (SSP 4). "Historical Trends" (SSP 2) will be considered thereafter and finally "Regional Rivalry" (SSP 3).

"Sustainability" (SSP 1). A demographic analysis (KC and Lutz 2017) found that achievement of SDG 4 (Quality Education) and SDG 5 (Gender Equality) would likely result in the global population of SSP 1 peaking mid-century and declining to seven billion by 2100. An economic analysis (Dellink *et al.* 2017) found that prioritizing convergence between developing and developed countries (a policy consistent with SDG 10, "Reduced Inequalities")

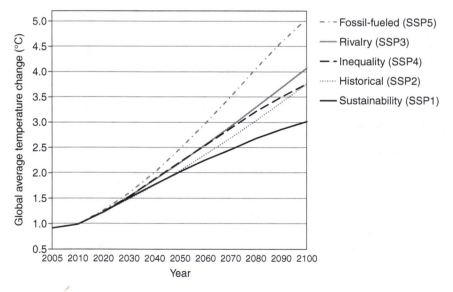

Figure 9.2 Projected global average temperature change across baseline versions of the Shared Socio-economic Pathways.

Source: this figure is based on the SSP database hosted by the IIASA Energy Program at https://tntcat.iiasa.ac.at/SspDb. In the absence of climate policy, all SSPs drive temperature change outcomes beyond the Paris target. As discussed in the chapter, SSP 3 cannot achieve the Paris target. For SSP 1, costs for achieving the Paris target are lowest, while costs for SSP 5, SSP 4, and SSP 2 could be double SSP 1. SSP 2 has a wide range of feasible carbon prices, either being similar to SSP 1 or exceeding SSP 4. Cost is a function of when global climate policies are implemented, the rate of technological change, and land-use emissions.

would increase the global income growth rate in the near term (furthering SDG 8, "Decent Work and Economic Growth"). Thereafter, even under moderate assumptions for the rate of income growth for the rest of the century, an eight-fold increase of income would be expected by 2100. An integrated analysis (van Vuuren *et al.* 2017b) found that global energy demand would increase slowly in this world, as efficiency improvements could match increases in demand. The biggest efficiency improvement comes from accomplishing SDG 7 (Affordable and Clean Energy), by rapidly phasing out traditional bioenergy use. Achievements in SDG 1 (No Poverty) and SDG 2 (Zero Hunger) increase global per capita food consumption. However, land use for crop and pasture land would still decrease due to: decreased food waste (SDG Target 12.3 for Responsible Consumption and Production), decreased demand for animal products in wealthy countries (resulting in a decreased material footprint for diets, i.e. SDG indicator 12.2.1), increased efficiencies in livestock systems, and rapid technological development that raises crop yields (e.g. Dewey 2018). Decreased use of land for food or biofuels increases natural land cover by hundreds of millions of hectares (furthering SDG 15, Protection and Restoration of Terrestrial Ecosystems). This also improves natural CO_2 removal from the atmosphere. Global air quality improves due to a shift away from fossil fuels and internal combustion engines (advancing SDG Target 11.6 to decrease adverse environmental impacts of cities).

Because of the above trends in efficiency improvements for clean energy and land use, even in the absence of additional climate policy, SSP 1 will result in fewer emissions in 2100 compared to today (a 35 per cent decrease). However, this decrease is still not enough to achieve the UNFCCC Paris target of limiting global average temperature increase to 2.0 degrees Celsius. As shown in Figure 9.2, on its own, SSP 1 achieves a temperature change of approximately three degrees Celsius. To achieve the Paris target, carbon pricing[6] would still be needed. SSP 1 yields the lowest possible carbon prices to achieve the Paris target. This is because lower emissions require less effort to bring the global average temperature change down further to 2.0 degrees Celsius. In SSP 1, carbon taxes are as low as 0.38 per cent of global GDP (Riahi *et al.* 2017, Supplementary Material). Van Vuuren *et al.* (2017b) note that prices on the lower end of the estimated range depend on international co-operation (SDG 16 for Strong Institutions and SDG 13 for Climate Action) that leads to early co-operative learning in order to settle on globally optimal climate policies as soon as possible.

"Fossil-fuelled Development" (SSP 5) bears the trappings of sustainable development but achieves them with much less concern for climate action (SDG 13). Similar to "Sustainability", by 2100, global population is under eight billion people for SSP 5. Throughout the century in this scenario, annualized growth rates for income remain above 2 per cent, leading to a 14-fold increase of income by 2100 (Dellink *et al.* 2017).[7] Its relationship to SDG 13 (Climate Action), however, could be characterized as discrepant. SSP 5 depicts a highly engineered future that may be consistent with many local or national governments having strategies for disaster risk reduction and climate adaptation

(SDG targets 13.1–13.2), education or training programs to implement adaptation development actions (SDG targets 13.3 and 13.b), and adequate support of the Green Climate Fund for adaptation programs. However, SSP 5 is also characterized by lack of concern with decreasing emissions (i.e. failures on SDG indicators 13.2.1 for operationalizing policies/strategies/plans for low greenhouse gas emissions, 12.C.1 for decreased fossil-fuel subsidies per unit GDP or as a proportion of total national expenditure on fossil fuels, and 9.4.1 for decreasing carbon intensity). There is also little interest in developing renewable energy technology, and no constraints on fossil fuels. In short, the question for SSP 5 is whether it may be possible to leave historically fossil-fuelled and energy-intensive economies primarily intact and adapt our way out of climate change impacts. From their analysis, Kriegler *et al.* (2017) verified that rapid economic growth and commitments to SDG 1 (No Poverty), SDG 8 (Decent Work and Economic Growth), and SDG 10 (Reduced Inequalities) could result in low income-shares being spent on food and energy globally, even when energy prices increase due to the depletion of conventional fossil resources (e.g. peak oil). However, in SSP 5, the use of unconventional fossil resources and technologies – such as tar sands and coal-to-liquids – results in the highest emissions profiles entertained for climate scenarios, corresponding to a temperature change greater than 4.0 degrees Celsius by 2100.

Numerous studies have warned that warming at this level would be intolerable (see e.g. The World Bank 2012), as climatic stressors would exceed coping mechanisms or adaptive capacity. Under such circumstances, economic growth and development would stall. Should the climate impacts of the SSP 5 world prove to be intolerable, Kriegler *et al.* (2017) found that aggressive carbon prices (ranging from $200–$470/tCO$_2$ post 2050) could likely bend the SSP 5 emissions trajectory down toward a temperature change of 2.0 degrees Celsius by 2100. This finding means that it may remain technically feasible to achieve the objective of the UNFCCC in SSP 5. This is because the SSP 5 world is technologically advanced, globalized (upholding SDG 16, Strong Institutions), and sufficiently wealthy. However, achieving the Paris target would come at high cost in comparison to SSP 1, which can achieve similar results at half the carbon price or less (Riahi *et al.* 2017). Moreover, this socio-economic pathway is a climate gamble. Not all integrated assessment models participating in the "Global Environmental Change" special issue could feasibly achieve the Paris target under socio-economic assumptions consistent with SSP 5. Thus, there is some scientific uncertainty whether SSP 5 hits a limit to climate change mitigation. In short, although SSP 5 may be a world where there is a chicken in every pot, a car in every driveway, and at least one long-haul vacation for every household per year, it is not truly sustainable development. By de-prioritizing mitigation action, it is not climate resilient. Thus, the costs of delaying emissions reductions accumulate and exceed the carbon prices of the climate-resilient pathway of "Sustainability" (SSP 1).

"Inequality" (SSP 4) juxtaposes socio-economic conditions that would lower mitigation challenges (especially with respect to energy use in modernized countries) but retains conditions keeping global adaptation challenges high. This scenario is characterized by insufficient commitment to SDG 10 (Reducing Inequalities), resulting in limited success in the poorest regions with SDG 4 (Quality Education), SDG 5 (Gender Equality), and possibly SDG 3 (Good Health and Well-being). For these reasons, global population is higher at approximately 9.5 billion people. Such outcomes may seem inconsistent *prima facie* with sustainable development; however, SSP 4 sounds a cautionary note for how the SDGs are pursued. In the MDGs, donors assisted recipient countries to achieve goals, whereas the SDGs are "one-world" goals where each country pursues the SDGs according to its ability (UN System Task Team on the Post-2015 UN Development Agenda 2013). Due to a lack of commitment to reducing inequality, SSP 4 describes a future where collective action is implemented perversely. "[T]he ability to achieve [SDGs] is regionally differentiated. [High-income regions] make progress toward eradicating hunger, improving air quality, and providing energy access. [Low-income regions], however, continue to struggle to meet these goals" (Calvin *et al.* 2017, 295).

This scenario is characterized by high inequality within countries. Income growth in developing countries becomes especially stunted, slowing overall annualized global growth of income to 0.7 per cent (Dellink *et al.* 2017). The lack of commitment to improving inequality also means that energy access for poorer populations is of low priority. This results in slower modernization of the final global energy mix (Bauer *et al.* 2017). In contrast to "Sustainability" (SSP 1), which depicts a rapid phaseout of traditional biofuel, in SSP 4, such fuels still comprise 35 per cent of energy use in the built environment in low-income regions by 2100 (Calvin *et al.* 2017). Meanwhile, high-income regions successfully use clean energy such as renewables or nuclear power (Calvin *et al.* 2017). Despite having a global population similar in size to "Historical Trends" (SSP 2), in SSP 4, there are relatively low increases in the market demand of crops and livestock products. This is because increases in population are primarily in low-income regions with low agricultural productivity and limited access to markets (Popp *et al.* 2017). Additionally, elites in SSP 4 are unconcerned with potential fluctuations in food prices driven by demand for biofuels, which competes with food for arable land. This results in a large expansion of cropland driven by demand for biofuels (Popp *et al.* 2017). Global land dynamics are such that high-income regions afforest for the purposes of natural CO_2 removal and possibly to improve environmental quality near their own borders (a locally focused implementation of SDG 15). Meanwhile, low-income regions deforest to grow biofuel and food (Calvin *et al.* 2017).

Due to the above socio-economic characteristics, SSP 4 achieves a temperature change of approximately 3.7 degrees Celsius. As shown in Figure 9.2, given the smaller radiative forcing compared to "Fossil-fuelled Development" (SSP 5), one might conclude that carbon prices for achieving the Paris temperature target should be lower. However, Calvin *et al.* (2017) estimated carbon prices higher than SSP 5, exceeding $2000/t by 2100. This happens because SSP 4 has the technological capacity to reduce emissions in the energy sector;

however, due to neglect of populations and institutions working in low-income regions, emissions from land use (i.e. from tropical deforestation, CH_4 emissions from livestock and rice cultivation, and N_2O from fertilized soils and manure as described by Popp *et al.* [2017]) are difficult to decrease.

SSP 4 demonstrates that success with the SDGs for the wealthiest countries and wealthiest populations within developing countries is not truly sustainable development. Additionally, if emissions reductions are focused on the global energy sector without commensurate attention to emissions from land use, land-use emissions could become the primary driver of climate change. Thus the uneven effort towards emissions reductions in SSP 4 will also cause carbon prices to exceed those of the climate-resilient pathway of "Sustainability" (SSP 1).

"Historical Trends" (SSP 2) is a future with continued progress on SDGs but without the major breakthroughs on Human Development achieved by "Fossil-fuelled Development", SSP 5, or clean energy technologies achieved by "Inequality", SSP 4. Similar to SSP 4, slower progress on SDG 4 (Quality Education) and SDG 5 (Gender Equality) lead global population to peak at 9.4 billion in 2070, then slowly decline. GDP growth is consistent with regional historical trends, and 40–70 years from now (the years 2060–2090), developing countries finally reach the average income levels of OECD countries observed today. Unfortunately, economic growth in Africa continues to lag due to modest progress on educational attainment, i.e. SDG 4 (Fricko *et al.* 2017). Slower progress on SDG 7 (Clean Energy Access) leads to moderate modernization of final energy use globally, with traditional biofuels phased out by 2080 (Fricko *et al.* 2017). The SSP 2 energy system favours neither clean energy nor fossil fuels. Thus, by 2050, up to 30 per cent of electricity worldwide could come from renewables (van Vuuren *et al.* 2017b); however, oil continues to dominate, even if it must be derived from unconventional sources such as tar sands (Fricko *et al.* 2017). In Asia, the Middle East, and Africa, coal fuels industrial development, doubling its volume by 2050 and tripling by 2100 (Bauer *et al.* 2017). The continued use of fossil fuels, under air-quality policies only moderately better than today, slows improvement in global air quality (van Vuuren *et al.* 2017b). Food demand overall increases moderately, with the highest shares and increases being in Asia (Popp *et al.* 2017). Agriculture systems in SSP 2 are more intensive than "Sustainability" (SSP 1), yet global population is higher, making conversion of natural land cover to productive land cover more likely (van Vuuren *et al.* 2017b). Fricko *et al.* (2017) projected a global increase of crop production by 2100 that is 84 per cent higher than levels in 2010, suggesting modest or no progress on SDG 15 (Protection and Restoration of Terrestrial Ecosystems).[8]

Because of the above trends, in the absence of additional climate policy, SSP 2 results in a doubling of emissions by 2100 (Fricko *et al.* 2017). On its own, SSP 2 achieves a temperature change of 3.5–4.0 degrees Celsius (Fricko *et al.* 2017; van Vuuren *et al.* 2017b). The Paris target remains achievable, but the range of feasible carbon prices is large. They could be as low as some analysed for "Sustainability" (SSP 1) or among the highest for scenarios that achieve the Paris target (Riahi *et al.* 2017). As noted for "Sustainability" (SSP 1), the lowest carbon prices likely depend upon early commitment to a global carbon price, which requires commitment to Strong Institutions (SDG 16). Higher carbon prices could be a reflection of slower rates of technological change in SSP 2 compared to "Sustainability" (SSP 1), "Fossil-fuelled Development" (SSP 5), and "Inequality" (SSP 4).

"Regional Rivalry" (SSP 3) is characterized by slowdowns or reversals in all historical trends for globalization and development. Due to lack of progress on SDG 4 (Quality Education), SDG 5 (Gender Equality), and probably SDG 3 (Good Health and Well-being), global population is high at 12 billion people (Fujimori *et al.* 2017; KC and Lutz 2017). Annualized growth of income is the lowest of all SSPs at 0.5 per cent, resulting in about a doubling in income levels by 2100 (Dellink *et al.* 2017). Saddled with the slowest modernization rate for the global energy mix, integrated assessments of SSP 3 found that energy access (i.e. electrification, SDG 7) in developing regions never catches up with developed regions by the end of the century (Bauer *et al.* 2017). Additionally, with trade being restricted due to national security concerns (including energy security), most of the global energy supply is coal, as it is the predominant domestic fuel in Asia (van Vuuren *et al.* 2017b). Due to the continued use of fossil fuels (especially coal) and lack of political appetite to improve existing air quality measures, there is no improvement in global air quality (Fujimori *et al.* 2017; van Vuuren *et al.* 2017b). The large global population puts more strain on the food system, leading to conversions of natural land cover to productive land cover that are similar to or more extensive than "Historical Trends" (SSP 2). In their integrated assessment, Fricko *et al.* (2017) found that almost twice as much additional land and irrigation water was needed for the global food system compared to "Historical Trends". Popp *et al.* (2017) found SSP 3 to have the highest conversion of natural areas to crop or pastureland, with this occurring mostly in the Middle East, Africa, and Latin America, which may negatively impact biodiversity (SDG 15). Moreover, in SSP 3, fewer people can afford food (lack of progress on SDG 1). Thus global per capita food consumption is lower than "Historical Trends" (van Vuuren *et al.* 2017b), indicating lack of progress on SDG 2 (Zero Hunger).

SSP 3 results in a temperature change of 3.5–4.5 degrees Celsius (van Vuuren *et al.* 2017b). Importantly, Bauer *et al.* (2017) find the Paris target to be unachievable due to weak near-term mitigation policy ambition, slow technological progress in the energy sector throughout the century, and higher demands for arable land to be used for food production due to the larger global population. Fujimori *et al.* (2017) noted that the higher level of agricultural emissions in SSP 3 are more difficult to reduce and mitigate, a challenge similar to that encountered by "Inequality" (SSP 4). Importantly, Fujimori *et al.* (2017, 275) note, "Net CO_2 emissions can be negative through the incorporation of ... [carbon capture and sequestration] or afforestation, whereas there are no equivalent countermeasures for non-CO_2 emissions [i.e., CH_4, N_2O]". To achieve the most ambitious emissions reduction feasible for SSP 3 by 2100 (a temperature change in the range of 2.0–2.5 degrees Celsius), carbon prices are still high at $1,120/tCO$_2$eq, with some analysts finding prices as high as $3416/tCO$_2$ (Fujimori *et al.* 2017). Early studies of climate impacts under the socio-economic characteristics of SSP 3 have already found an increased number of people at risk of hunger (failure achieving SDG 2 [Zero Hunger]; see Hasegawa *et al.* [2015]) and water stress (failure achieving SDG 6 [Clean Water and Sanitation]; see Hanasaki *et al.* [2013]).

Discussion and conclusion

This chapter introduces and summarizes a two-part scenario framework for climate change research that makes the importance of development abundantly clear for understanding climate risk as well as the level of challenge for climate policy. The two-part framework separates socio-economic conditions (non-climatic factors) from climatic conditions. This is scientifically justified because of new thinking around the nature of climate risk as well as findings that different socio-economic conditions can give rise to similar emissions profiles. The latest socio-economic scenarios, SSPs, are sufficiently detailed that the level of progress on most SDGs can be discerned.

Qualitatively, the SSPs describe five alternative worlds:

- A best-case scenario, where all SDGs are achieved or nearly achieved ("Sustainability", SSP 1);
- A continuation of historical trends (SSP 2), where SDGs may or may not be achieved;
- A worst-case scenario, where there is a slowdown or reversal of historical trends in globalization and development due to concerns with national security ("Regional Rivalry", SSP 3);
- A "second-best" world describing inequitable achievement of the SDGs, where wealthier countries and communities achieve most of them (including clean energy), but the poorest communities within and across countries are left behind ("Inequality", SSP 4);
- Another "second-best" world, where most SDGs are achieved but at the expense of timely mitigation policy ("Fossil-fuelled Development", SSP 5).

Recent research has quantified the SSPs and found that "Regional Rivalry" (SSP 3) cannot achieve the UNFCCC Paris Agreement target limiting global average temperature change to 2.0 degrees Celsius by 2100. "Fossil-fuelled Development" (SSP 5) also runs the risk of possibly missing the Paris target in spite of future generations being much richer and much more technologically capable than today. All other SSPs may reach the Paris target, with "Sustainability" (SSP 1) being the least expensive and "Historical Trends" (SSP 2) potentially being the most expensive.

A close look at the drivers of the outcomes of the ideal case ("Sustainability") versus alternative cases suggests that particular SDGs should be prioritized to achieve multiple SDGs and climate-resilient outcomes.

- Investments to improve Human Development (especially **SDG 4 [Quality Education]** and **SDG 5 [Gender Equality]**) have important demographic effects for achieving SDG 8 (Decent Work and Economic Growth), SDG 1 (No Poverty), and SDG 2 (Zero Hunger) (see also Lutz [2017]). This is apparent when comparing the outcomes of "Sustainability" (SSP 1) to "Historical Trends" (SSP 2), "Inequality" (SSP 4), and "Regional Rivalry" (SSP 3).
- Climate-resilient development requires both clean energy and sustainable food systems. In their comparison of findings across SSPs 1–3, van Vuuren *et al.* (2017b) conclude that energy and resource efficiency improvements (SDG 7 [Affordable and Clean Energy], and SDG 12 [Responsible Consumption and Production]) explain much of the long-term benefits for reduced mitigation challenges for "Sustainability" (SSP 1). SSP 1 has emissions that are 35 per cent lower than today, and reductions in both food waste and meat consumption lower demand for crop- and pastureland. By preserving the most land in its natural state, "Sustainability" yields the greatest benefits for terrestrial biodiversity (SDG 15). Other SSPs require more natural areas be converted to productive land, with the high-population and materially intensive scenario of "Regional Rivalry" (SSP 3) requiring the most. Both "Inequality" (SSP 4) and "Regional Rivalry" (SSP 3) show that high land-use emissions (i.e. methane and nitrous oxide) are more difficult to offset than carbon dioxide emissions from the energy sector.
- For "Regional Rivalry" (SSP 3), its focus on national security – to the detriment of strong international institutions (SDG 16) – delays the build-up of domestic and international capacity for climate action for too long. SSP 3 is the only SSP that unequivocally cannot achieve the UNFCCC Paris target.[9] Its emissions profile is not terribly different from SSPs 2–4. This shows that its weak international institutions and partnerships (i.e. failures for SDG 16, SDG 17) contribute to its downfall. Lower carbon prices for "Historical Trends" (SSP 2) also depend upon early cooperative climate action, with carbon taxes going into effect this decade (2020).

Prior to our awareness of the threat that climate change poses, policy priorities like the SDGs could be mere aspirations or "the right thing to do". Now, climate change imposes deadlines on multiple issues acknowledged in the SDGs: human development, energy transitions, and sustainable food systems, and there are penalties for delaying action much further beyond the 2020s. No longer can we afford a *laissez-faire* attitude toward free markets and autonomous technological change to wait-and-see whether they will work out the problems of global change eventually. It is time to take more responsibility for the future and to do so with a co-operative commitment to sustainable development for all communities in all countries.

Acknowledgements

Inspiration for this chapter came from the author's participation in a 2013 inter-disciplinary workshop hosted at the Balsillie School of International Affairs entitled "What is Climate Change, and What Should We Do about It?"

Notes

1 According to the US National Oceanic and Atmospheric Administration, the concentration of carbon dioxide in the atmosphere is just over 410 ppm. For the Fifth Assessment Report, the IPCC estimated that global average temperature increased 0.85 degrees Celsius since 1880. It may very well be the case that at higher levels of emissions (and therefore higher levels of committed climate change), changes in climate may overwhelm socio-economic abilities to successfully adapt.
2 An example of a study that teases apart the relative contributions of different uncertainties is "The Potential to Narrow Uncertainty in Regional Climate Predictions" by Ed Hawkins and Rowan Sutton, published in the *Bulletin of the American Meteorological Society* in 2009. This study was performed with climate models; however, in theory, a similar study could be done with a sectorial model contrasting different sets of assumptions for climate and socio-economic scenarios.
3 SDG 13 is an explicit call for climate policy, which the SSPs purposely avoid including for research design purposes. However, socio-economic outcomes consistent with effective climate policy, such as constraints on fossil fuel extraction, are entailed in the SSPs. Socio-economic outcomes consistent with SDG 13 could be the result of other drivers, such as concerns over environmental degradation from drilling or oil spills.
4 Kriegler *et al.* (2017) explained that "Fossil-fuelled Development" purposely explores the combination of the strongest economic growth among SSPs, strong reliance on fossil fuels, and energy intensive consumption to consider a future with very large challenges to mitigation. These assumptions are not intended to posit that high carbon and resource intensity are necessary for high economic growth.
5 What level of climate change natural systems can adapt to matters as well, as many socio-economic systems depend upon the health of the natural environment for ecosystem services such as pollination and water purification.
6 Van Vuuren *et al.* (2017b) note that carbon prices should be seen as indicators of effort to reach an emissions reduction target. It's possible that some governments would prefer a regulatory approach to reducing emissions rather than taxes; however, the former may introduce inefficiencies to the economy and to technological innovation that carbon taxes would avoid.
7 Kriegler *et al.* (2017) caution that SSP 5 should not be interpreted to mean that fossil fuel use is a precondition for strong economic growth. SSP 5 was designed to explore the implications of high challenges to mitigation, which would be the case if high economic growth were combined with high fossil-fuel use.
8 As noted by Popp *et al.* (2017), already approximately 40 per cent of the terrestrial surface is under agricultural use as cropland or pasture.
9 Riahi *et al.* (2017) summarized climate policy assumptions for each SSP that would be consistent with its socio-economic conditions and policy milieu. Climate policy for SSP 3 is characterized by slow global co-operation, where countries delay joining a global policy regime until 2030–2050, and mechanisms for decreasing land-use emissions (e.g. REDD) are very limited.

References

Bauer, N., K. Calvin, J. Emmerling, O. Fricko, S. Fujimori, J. Hilaire, J. Eom, *et al.* 2017. "Shared Socio-Economic Pathways of the Energy Sector: Quantifying the Narratives". *Global Environmental Change* 42: 316–330.

Birkmann, J., S. L. Cutter, D. S. Rothman, T. Welle, M. Garschagen, B. van Ruijven, B. O'Neill, *et al.* 2013. "Scenarios for Vulnerability: Opportunities and Constraints in the Context of Climate Change and Disaster Risk". *Climatic Change* 133 (1): 53–68.

Calvin, K., B. Bond-Lamberty, L. Clarke, J. Edmonds, J. Eom, C. Hartin, S. Kim, *et al.* 2017. "The SSP4: A World of Deepening Inequality". *Global Environmental Change* 42: 284–296.

Dellink, R., J. Chateau, E. Lanzi, and B. Magné. 2017. "Long-term Economic Growth Projections in the Shared Socioeconomic Pathways". *Global Environmental Change* 42: 200–214.

Denton, F., T. J. Wilbanks, A. C. Abeysinghe, I. Burton, Q. Gao, M. C. Lemos, T. Masui, K. L. O'Brien, and K. Warner. 2014. "Climate-Resilient Pathways: Adaptation, Mitigation, and Sustainable Development". In *Climate Change 2014: Impacts, Adaptation, and Vulnerability*, edited by C. B. Field, V. R. Barros, D. J. Dokken, K. J. Mach, M. D. Mastrandrea, T. E. Bilir, M. Chatterjee, *et al.*, 1101–1131. New York: Cambridge University Press.

Dewey, C. 2018. "The Future of Food: Scientists have Found a Fast and Cheap Way to Edit your Edibles' DNA". *Washington Post*, 11 August.

Fricko, O., P. Havlik, J. Rogelj, Z. Klimont, M. Gusti, N. Johnson, P. Kolp, *et al.* 2017. "The Marker Quantification of the Shared Socioeconomic Pathway 2: A Middle-of-the-Road Scenario for the 21st Century". *Global Environmental Change* 42: 251–267.

Fujimori, S., T. Hasegawa, T. Masui, K. Takahashi, D. S. Herran, H. Dai, Y. Hijioka, and M. Kainuma. 2017. "SSP3: AIM Implementation of Shared Socioeconomic Pathways". *Global Environmental Change* 42: 268–283.

Hanasaki, N., S. Fujimori, T. Yamamoto, S. Yoshikawa, Y. Masaki, Y. Hijioka, M. Kainuma, *et al.* 2013. "A Global Water Scarcity Assessment under Shared Socio-economic Pathways – Part 2: Water Availability and Scarcity". *Hydrology Earth System Sciences* 17: 2393–2413.

Hasegawa, T., S. Fujimori, Y. Shin, A. Tanaka, K. Takahashi, and M. Toshihiko. 2015. "Consequence of Climate Mitigation on the Risk of Hunger". *Environmental Science and Technology* 49 (12): 7245–7523.

Hawkins, E., and R. Sutton 2009. The Potential to Narrow Uncertainty in Regional Climate Predictions". *Bulletin of the American Meteorological Society*. August. 1095–1107.

IIASA. 2016. "SSP Database". Accessed 11 May 2018. https://tntcat.iiasa.ac.at/SspDb/dsd?Action=htmlpage&page=about.

IPCC. 2012. *Managing the Risks of Extreme Events and Disasters to Advance Climate Change Adaptation*. A Special Report of Working Groups I and II of the Intergovernmental Panel on Climate Change. New York: Cambridge University Press.

IPCC. 2014. *Climate Change 2014: Synthesis Report*. Contribution of Working Groups I, II and III to the Fifth Assessment Report of the Intergovernmental Panel on Climate Change. Geneva: IPCC.

KC, S., and W. Lutz. 2017. "The Human Core of the Shared Socioeconomic Pathways: Population Scenarios by Age, Sex and Level of Education for All Countries to 2100". *Global Environmental Change* 42: 181–192.

Kriegler, E., N. Bauer, A. Popp, F. Humpenöder, M. Leimbach, J. Strefler, L. Baumstark, et al. 2017. "Fossil-fueled Development (SSP5): An Energy and Resource Intensive Scenario for the 21st Century". *Global Environmental Change* 42: 297–315.

Lutz, W. 2017. "Global Sustainable Development Priorities 500 y after Luther: Sola schola et sanitate". *Proceedings of the National Academy of Sciences of the United States of America* 114 (27): 6904–6913.

Nakicenovic, N., R. J. Lempert, and A. C. Janetos. 2014. "A Framework for the Development of New Socio-economic Scenarios for Climate Change Research: Introductory Essay". *Climatic Change* 122 (3): 351–361.

O'Neill, B. C., T. R. Carter, K. L. Ebi, J. Edmonds, S. Hallegatte, E. Kemp-Benedict, E. Kriegler, et al. 2012. *Meeting Report of the Workshop on The [sic] Nature and Use of New Socioeconomic Pathways for Climate Change Research*. Boulder, CO, National Center for Atmospheric Research.

O'Neill, B. C., E. Kriegler, K. L. Ebi, E. Kemp-Benedict, K. Riahi, D. S. Rothman, B. J. van Ruijven, et al. 2017. "The Roads Ahead: Narratives for Shared Socioeconomic Pathways Describing World Futures in the 21st Century". *Global Environmental Change* 42: 169–180.

Popp, A., K. Calvin, S. Fujimori, P. Havlik, F. Humpenöder, E. Stehfest, B. L. Bodirsky, et al. 2017. "Land-use Futures in the Shared Socio-economic Pathways". *Global Environmental Change* 42: 331–345.

Riahi, K., D. P. van Vuuren, E. Kriegler, J. Edmonds, B. C. O'Neill, S. Fujimori, N. Bauer, et al. 2017. "The Shared Socioeconomic Pathways and their Energy, Land Use, and Greenhouse Gas Emissions Implications: An Overview". *Global Environmental Change* 42: 153–168.

Sanwal, M. 2012. "Rio + 20, Climate Change and Development: The Evolution of Sustainable Development (1972–2012)". *Climate and Development* 4 (2): 157–166.

Schweizer, V. J., and B. C. O'Neill. 2014. "Systematic Construction of Global Socio-economic Pathways Using Internally Consistent Element Combinations". *Climatic Change* 122 (3): 431–445.

Steffen, W., K. Richardson, J. Rockström, S. E. Cornell, I. Fetzer, E. M. Bennett, R. Biggs, et al. 2015. "Planetary Boundaries: Guiding Human Development on a Changing Planet". *Science* 347 (6223), 1259855-1-1259855-10.

The World Bank. 2012. *Turn Down the Heat: Why a 4°C Warmer World Must be Avoided*. (No. 74455). The World Bank.

UN System Task Team on the Post-2015 UN Development Agenda. 2013. *A Renewed Global Partnership for Development*. New York, March. Accessed 31 October 2018. https://sustainabledevelopment.un.org/content/documents/833glob_dev_rep_2013.pdf.

United Nations. 1992. *United Nations Framework Convention on Climate Change*. FCCC/INFORMAL/84. Accessed 26 October 2019. https://unfccc.int/resource/docs/convkp/conveng.pdf.

van Vuuren, D. P., J. Edmonds, M. Kainuma, K. Riahi, A. Thomson, K. Hibbard, G. C. Hurtt, et al. 2011. "The Representative Concentration Pathways: An Overview". *Climatic Change* 109 (5): 5–31.

van Vuuren, D. P., K. Riahi, R. Moss, J. Edmonds, A. Thomson, N. Nakicenovic, T. Kram, et al. 2012. "A Proposal for a New Scenario Framework to Support Research and Assessment in Different Climate Research Communities". *Global Environmental Change* 22 (1): 21–35.

van Vuuren, D. P., E. Kriegler, B. C. O'Neill, K. L. Ebi, K. Riahi, T. R. Carter, J. Edmonds, et al. 2014. "A New Scenario Framework for Climate Change Research: Scenario Matrix Architecture". *Climatic Change* 122 (3): 373–386.

van Vuuren, D. P., K. Riahi, K. Calvin, R. Dellink, J. Emmerling, S. Fujimori, S. Kc, E. Kriegler, and B. O'Neill. 2017a. "The Shared Socio-economic Pathways: Trajectories for Human Development and Global Environmental Change". *Global Environmental Change* 42: 148–152.

van Vuuren, D. P., E. Stehfest, D. E. H. J. Gernaat, J. C. Doelman, M. van den Berg, M. Harmsen, H. S. de Boer, *et al.* 2017b. "Energy, Land-use and Greenhouse Gas Emissions Trajectories under a Green Growth Paradigm". *Global Environmental Change* 42: 237–250.

10 Religion and the Sustainable Development Goals

Paul Freston

Introduction: "religion and the SDGs" as homologous to "religion and politics"

A book on the Sustainable Development Goals (SDGs) can ill-afford to overlook religion. This claim is not based solely on the oft quoted estimate that over four-fifths of the world's population is affiliated to a religion (Pew 2012). My starting-point in this chapter will be that the relationship between religion and the SDGs is homologous to the general relationship between religion and politics. There are three common errors about religion and global politics. The first is to ignore religion as irrelevant, either because we simply do not see it (not having been trained to see it, or worse, having been trained not to see it). The second is to treat it in a reductionist fashion as merely epiphenomenal; we see it, but only as pointing to other, more "real" factors. And the third is to treat it as indeed very important and not epiphenomenal, but as all the same and all bad; as uniform, and uniformly negative in its effects. In other words, in rhyming fashion (and using first an old English word meaning to withdraw or subtract), the three errors, respectively, *subduce, reduce* or *traduce* religion.

We should not, of course, fall into the opposite errors of seeing religion as *everywhere, all important* and *always positive*. Instead, along the lines of Max Weber's ([1904–1905] 2005) approach to the "economic ethic of the world religions", we should see it as always a *possible autonomous factor in world affairs (pace* the subducers), able to have an independent influence on the course of events (*pace* the reducers), whether for good or ill, being (like any other human phenomenon) profoundly *ambivalent (pace* the traducers). In a globalizing world, social science must be capable of taking religion seriously as a possible autonomous actor in public life.

In addition, we should recognize the multifaceted nature of religion in public life. Religion functions as ideas (about the world, the good society, rulers); it also functions as a source of identity (personal, group, national, global); it may be constituted in influential organizations; it can function as a tremendously powerful mobilizing force; it is a former of extensive networks; it is a provider of services, both to its own adepts and beyond; it is a highly influential source of ethics and values; it can function as a school (for learning to speak in public, to organize, to lead); and it has often, historically, been a template for institutional models (in trade unionism, in education, in handling diversity, etc.).

Sociologically, we should remember that the public effect of religious groups cannot be directly deduced from that group's theology. Theological ideas are just one influence among others that have to be taken into account. In the space between religious doctrine and public practice, many other factors come into play: from within the religious field itself, and from the socio-political context. Navigating between the extremes of essentialism and contextualism, we recognize that religious traditions are not univocal or immutable, and there is always diversity and evolution; nevertheless, each religious tradition develops within a certain socio-logic, with certain constraints and prevailing tendencies.

History: "religion and the SDGs" as an extension of "religion and development"

In her Introduction to the *Routledge Handbook of Religions and Global Development*, Emma Tomalin (2015, 1) says that "significant research" on this relationship is relatively recent; in the past decade or so, there has been a "noticeable shift within some areas of international development policy, practice and research to include religion as a relevant factor". Or, as Susan Holman (2015, 8) specifies in relation to the area of health, "especially since 2001 the role of religion in global health has become a topic for serious dialogue"; a phrase which reveals the tragicomic reality that religiously inspired terrorism (the 9/11 attack) was central in provoking serious engagement between such clearly related areas as religion and health.

The relationship between religion and the modern development project can be seen as a sub-set of the larger trajectory of Western assumptions regarding secularization and privatization of religion, but with a time-lag.

Marginalized by modernization theory (and by Marxism), religion was for long peripheral in the development arena. Even Max Weber's thesis on the historical role of the "Protestant ethic" and his explorations of the "economic ethic" of the other world religions do not claim any lasting role for religion in general or Protestant Christianity in particular. If we locate the origins of the modern development project in the post-war era, it was characterized by the assumption that religion was largely opposed to development and that as societies modernized, they would become less religious (Tomalin 2015, 1). Simplistic theories of modernization and secularization, and the early modern construction of a (containable) religious sphere, thus underpinned the "amnesia" which impeded development from acknowledging its religious (and missionary, and colonial) roots, notwithstanding its own "evangelical spirit" (Tomalin 2015, 4). Religious activists and development actors were too similar and too historically connected for comfort, hence the need to stress difference and separation. Nevertheless, as Ryan says (2012, 19), the development paradigm was amply embraced by the churches; the papal encyclical *Populorum Progressio*, of 1967, averred that "the progressive development of peoples is an object of deep interest and concern to the Church".

By the turn of the century, however, talk of "modernization" had been in large part replaced by "globalization", which (as Beyer [2006, 18–19] explains) cannot blithely be used as a successor term with impunity, since the old concept of a unitary modernity has been (implicitly, at least) dethroned either by Eisenstadt's concept of "multiple modernities" or by Huntington's "clash of civilizations". Religion and modernity might, after all, be related in unsuspected ways. Perhaps the first sign of what was coming could be found in a special 1980 issue of "World Development", in which Wilber and Jameson reflected on the Iranian Revolution's demonstration of how disastrous the consequences could be when development and the moral base of a society were sundered (Deneulin and Bano 2009, 38–39). But Iran was not the only sign of the times stemming from the fateful year of 1979; events as far afield as China, the Middle East, Poland, the UK, the US and Latin America also pointed to a new visibility of religion in public life, including the neoliberal turn which, in both developed and developing contexts, left space for religious actors to (re)enter social service provision.

Combined with the "human development" approach's recognition of development as a value-laden enterprise, this new reality led (with something of a time-lag in comparison to other areas of public life) to significant initiatives to rethink the role of religion. A prominent early example was the World Faiths Development Dialogue, initiated in 1998 by the president of the World Bank and the Archbishop of Canterbury. Then came significantly increased funding to so-called faith-based organizations (FBOs) by organs such as the United States Agency for International Development (USAID) and the UK's Department for International Development (DFID), accompanied by attempts at guidelines for engagement. In 2009, a United Nations Inter-Agency Task Force on Engaging with Faith-Based Organizations for Sustainable Development was formed.

Secularists were, of course, critical of the new trend, but also some sympathizers: a "pernicious instrumentalism" (Tomalin 2015, 3) led to choosing which religious groups to engage (the "hierarchical, carefully codified book religions" whose organizations resembled those of secular NGOs and which tended to express their faith merely "passively"), and to engaging religion only where it seemed instrumental to achieving the ends pre-determined by non-religious actors (Deacon and Tomalin 2015, 76). Adding a fourth rhyming term to the three mentioned above (although this one referring to a sin of practitioners rather than academics), they *seduce* religion, attracting it merely in order to instrumentalize it. In Thomas' (2005, 232) scathing evaluation, the "secular missionaries" of development agencies, if they regard religion as more than part of the problem, have usually reduced it to "thin" practices appealed to in terms of Enlightenment rationality, supporting a narrow range of local NGOs that fit secular utilitarian concepts; or at most they accept the collaboration of faith-based organizations as long as they do not proselytize or try to influence the *content* of development. Others only imagine a possible role for religion in developing countries as an "authentic" alternative to the modernization paradigm (Selinger 2004, 531), as a "revolt against the West" in which local cultural "authenticity" rivals "development" for primacy in political aspirations (Thomas 2005, 42). There is a need to take religious pluralism seriously, not as a post-modern celebration of diversity which argues that any normative truth claims are to be censured as divisive, but by accepting that real pluralism includes clashing normative claims.

In short, while much has changed (it is now "common for heads of development institutions to hold publicised meetings with religious leaders" [Olivier 2016]; and in 2016 the German government launched a network called the International Partnership on Religion and Sustainable Development), fears (among religious groups) of exploitation and of subordination to a purely secular framing of development issues remain. Ager and Ager complain of the continued predominance of a secular way of thinking, "peculiarly lacking in self-awareness of its origins and the weakness of its claims to universality", resulting in an "imposition of Western dualism ... cleaving worldly actions from spiritual sentiments" (2016, 104).

Although focusing on Europe rather than the developing world, Bäckström and Davie's (2010) work on welfare and religion highlights some intriguing points. First, that throughout Europe the role of the churches is increasing, both as providers of welfare and as a critical voice pointing out deficiencies in the system. Second, that in both welfare and religion, women are disproportionately present at the point of delivery but under-represented in management. Third, that there are significant differences between the various state churches on welfare: Lutherans welcome a strong state welfare sector in the name of their "two kingdoms" theology; Catholics are wary of state intrusion into social sectors they see as central to their identity; and the Reformed also resist state intrusion, but in the name of individualism and self-reliance as supreme virtues. Thus, religion has been an important independent variable in the evolution of diverse European welfare systems. And last, given the increasing marketization of care, the traditional European churches debate whether to participate in this or to resist it on principle; there is thus a tension between their "practical" and their "prophetic" functions (Bäckström and Davie 2010, 6).

Classification: "religion and the SDGs" through the rubric of FBOs – dubious categorization and insufficient evidence?

The FBO category has become central for development's engagement with religion, yet it is highly problematic. Clarke and Jennings (2007, 3) define an FBO as an organization that "derives inspiration and guidance for its activities from the teaching and principles of the faith or from a particular interpretation of the faith"; but that, complains Hefferan (2015, 42) is far too broad. Occhipinti (2015) agrees; the FBO category is often taken for granted, despite enormous variation. It can include places of worship or informal and local organizations, as well as formally registered organizations that may have an international reach. But they not only vary immensely in size, but also in the faith tradition which influences them, their approach to development, and the way they conceive of a dignified life. In addition, local actors have been known to manoeuvre strategically, based on whether they see it as beneficial or not to be named as faith-based (Olivier 2016); and, in any case, it is not easy to distinguish between faith-based and non-faith-based organizations in contexts where religion permeates all of life. Even where it does not, the boundary may be blurred. Habitat for Humanity focuses on the elimination of poverty housing "as a demonstration of the love and teaching of Jesus Christ" (Berkley Center 2007, 9). Oxfam and Amnesty International were founded by religious people, and Greenpeace began in a church basement inspired by the Quaker tradition of "bearing witness"; yet these are not normally regarded as FBOs.

The supposed advantages of FBOs are, says Occhipinti (2015), organizational (strong grassroots links, often in remote areas; long-term presence; local knowledge; local trustworthiness; good networks for fundraising) and motivational (commitment, enthusiasm, honesty). Freeman, talking of Pentecostals in Africa, claims they are often more effective change agents than development NGOs, because they are home-grown, focus on transforming subjectivities, create moral legitimacy for behaviours that clash with local values, and are often very democratic and thus seen as "moral and meaningful institutions" (2012, 24–26). They contrast, she says, with much post-developmental thinking which tends to look for "cultural essentials" and "authentic" cultural patterns with which Pentecostals do not fit; but Pentecostals also challenge modernist ideas of the "disenchantment of the world" (Dijk 2012, 103).

Others are less convinced about the value of FBOs. Some are unprepared to "scale up", lacking infrastructure, experience, and staff (Occhipinti 2015, 333). They may have problematic stances on gender equality, sexual and reproductive health, and religious tolerance. Karam (2015) complains that they may be stoking strife and undermining the basis of multilateralism. In extreme cases, they may even be funding terrorism (Marshall 2015, 383). Bompani (2015, 107–108) talks of the boycott by religious leaders in northern Nigeria of the Global Polio Eradication Initiative in 2003; the Supreme Council for Sharia in Nigeria asserted that the vaccines had been contaminated by Western governments as part of a plot to reduce Muslim populations worldwide. However, the case is also notable for the intervention of the Organization of the Islamic Conference, which pressured Islamic countries to make greater efforts against polio and wisely encouraged them to procure vaccines from Muslim nations. And as a UN report recognized, "faith-based organizations are themselves divided on … sexual and reproductive health and reproductive rights" (Redo 2015). Nevertheless, as Marshall (2011, 45) warns, "development practitioners surely need to avoid associating with advocates of violence and with bitter critics of other faiths". Similarly, Deneulin and Bano (2009, 105) mention four sticking-points for development practitioners: religious decrees which prohibit women from working; religious groups which consider each other as enemies to be converted or killed; denial of contraception to women with HIV-positive partners; and religious education which teaches children to hate those of other religions.

As for FBOs' supposed advantages over secular NGOs, Occhipinti stresses that they "have not been empirically established ... [Such claims] need to be treated sceptically. The recent wave of enthusiasm for FBOs may be founded more in changing global political realities ... than on any empirical evidence" (2015, 342). While some scholars talk of "compelling" and "increasingly robust" evidence for the added value of FBOs (Olivier 2016; Duff *et al.* 2016), others complain of methodological dilemmas or sheer lack of data, and conclude that "overall, the empirical evidence is mixed, inconclusive and sometimes contradictory" (Basedau, Gobien, and Prediger 2017, 33). Bäckström and Davie are similarly unclear about the European welfare context: "the data reveal different voices" (2010, 186). Supporters of FBOs claim "their services are more personal and are based not only on a genuine love of neighbour but on a thorough knowledge of the territory", whereas detractors are concerned "that giving of services might become conditional on church attendance or certain behavioural codes" (186–187).

But there is one more perceived disadvantage which hangs over many religious actors in development work: that of a presumed propensity to proselytize.

Proselytization: "religion and the SDGs" as an extension of the missionary spirit

Philip Fountain asserts baldly that "development has a problem with proselytization"; its intermixing with development work is regarded as "illegitimate, coercive and dangerous", requiring discursive policing and legal measures by major donor governments to restrain it (2015, 80). There are several problems with this, Fountain claims. It assumes that proselytism is an exclusively religious affair, as if similar sorts of advocacy did not exist elsewhere. The Red Cross Code of Conduct says: "aid will not be used to further a particular political or religious standpoint" (ICRC n/d, 3). Critiques of both development and humanitarianism have highlighted the impossibility of engaging in non-political activity, but a parallel critique about the bracketing of religion has hardly begun. Entities are allowed to "espouse" but not to "further" religious positions; where the difference lies exactly is unclear (Fountain 2015, 82). Above all, the underlying assumption is that it is possible to locate religion somewhere and not elsewhere: i.e. in the "religious sphere". Secularism is a redefinition of religion as locatable and boundable, as ahistorical and transcultural, and therefore available for policing (2015, 84).

> The dominant narrative of secularism imagines the secular as ... not itself a tradition ... but rather ... the default position of normality.... Secular development organizations pursue 'the good' without exposing their morality to the critique of parochialism ... [imagining] their values, ethics and practices as valid and transculturally applicable.
>
> (Fountain 2013, 22)

Thus, we might add, claiming the morality and universality they now deny to their precursors, the missionaries.

Why is this so insisted upon? Because, says Fountain (2015, 84), "proselytization is a polluting threat to the purity of the secular development enterprise ... [endangering] its ability to stand above cultural traditions". It is necessary because both development and humanitarianism are so profoundly indebted to particular religious histories, including a missionary heritage. Fountain cites Kroessin and Mohamed to the effect that "at an abstract level of analysis, all humanitarian and development actors ought ... to be regarded as 'missionaries'" (Fountain 2015, 89). They are underpinned by a set of values that drive them to promote social change and to shape the world in their image. Rather than defining proselytizing as inherently religious, we should define it as "intentional moral practices of transformative interventions aimed at reworking the social practices of others" (89). And this redefinition, Fountain concludes, is not in order to cast aside moral critique, but to force a new debate about modes of development, in which, "no longer concealed behind the pretence of secular neutrality, all development projects can and should be analysed as value-laden initiatives generated from within particular traditions" (91).

Progress: "religion and the SDGs" as an improvement on "religion and the MDGs"

The SDGs have generally been seen by religious actors as an improvement on the Millennium Development Goals (MDGs). For one thing, their elaboration was more inclusive (Frecheville 2015). They are "indivisible" (Frecheville 2015), underpinned by a "holistic understanding of development" (Deneulin and Zampini-Davies 2017). They recognize the importance of human dignity and people-centred economies (Frecheville 2015). Unlike the MDGs, which measured progress against blunt averages that masked growing gaps within and between countries, the SDGs are "multi-sectoral, rights-based and people-centred"; the "strictly secular framework for development gives way to greater inclusiveness" (Sidibé 2016, 1). The director of the International Social Justice Commission of the Salvation Army enthuses: "the aspirational holistic SDGs reflect the ambitions of those with religious faith ... [They are] God's agenda" (Santis 2016). The General Secretary of the World Association for Christian Communication agrees: "love thy neighbour, the Golden Rule ... governs the essence of the SDGs" (Santis 2016).

These positive Christian reactions are matched by those of inter-religious groups. In September 2015, leaders of 24 religious traditions from around the world met in Bristol to show their support for the 2030 Agenda for Sustainable Development. Even the criticisms voiced were along the lines of "nevertheless". One Bahai representative, for example, said: "in Agenda 2030 words like self-lessness, sacrifice, love, compassion, duty, generosity and charity are entirely absent", and the tone of the UN-FBO dialogue can be blunt at times; nevertheless, they still need to work together (Erasmus 2015).

In the Muslim world, the Islamic Leaders Climate Change Declaration (2015) affirmed that:

> We face the distinct possibility that our species, chosen to be God's care-taker (Khalifa) of the Earth, could be responsible for ending life as we know it on our planet.... We call on other faith groups to join us in collaboration, co-operation and friendly competition in this endeavour.

The Islamic Development Bank had already declared the MDGs "compatible with the Bank's 2020 Vision", especially in view of the fact that 40 per cent of the people living in absolute poverty worldwide were in its member-countries (UN Chronicle 2008). For Noor and Pickup (2017), there are "striking commonalities between the SDGs and zakat", the Islamic obligation to almsgiving. But they lament that, despite being one of the "largest forms of wealth transfer to the poor in existence", zakat organizations "have been overlooked by development organizations as ... source of finance".

The reference to "friendly competition" in the Islamic statement above encounters a more sombre echo in the 2017 conference of SAMVAD, the Global Initiative for Conflict Avoidance and Environmental Consciousness. SAMVAD is a joint initiative of Indian Prime Minister Narendra Modi and Japanese Prime Minister Shinzo Abe, "to adopt principles of Hinduism and Buddhism to address issues threatening human civilization ... to create a more tolerant, liberal and accommodative world" (an interesting stance from the leader of the Hindu nationalist BJP). Both Hindu and Buddhist speakers "repeatedly referred to problems in 'Abrahamic' scriptures and their lack of tolerance". Many Hindu speakers "spoke about how they have rejected old Hindu scriptures that speak of caste and 'untouchability' ... [and] they suggested Muslims in particular need to reject some of their ... scriptures that may preach exclusivity". But this conference was being held in Myanmar, and the Chief Convener of the Islamic Centre of Myanmar (coming from a situation where part of the Muslim minority is actively persecuted by the state and civil society elements, egged on by some Buddhist leaders) gave as good as he got:

> if anyone alleged that other religions are false or labelled the adherents ... as heathens ... dialogue would certainly be unproductive. It will be equally counterproductive to brand any religion, be it Abrahamic or not, as doctrinally intolerant and consisting in exhortation to religious violence or ... as not being ecologically friendly.

In every religion, "people ... hijack religion to suit their vested interests and hidden agendas". Nevertheless, the Chief Minister of Uttar Pradesh, a controversial Hindu priest-turned-politician, claimed Indian philosophies "do not thrust their points of view on others". He praised Buddhism, in what would seem to be an implicit critique of "Abrahamic" religions (Seneviratne 2017). Far from "friendly competition", therefore, this 2017 conference points in the direction of the SDGs as a renewed scene of inter-religious one-upmanship and even mudslinging. On the other hand, more positively, both the Hindu Declaration on Climate Change (2015) and the Buddhist Declaration on Climate Change (2015) talk of the elimination of fossil fuels and changing of consumption patterns in the light of the challenge of the SDGs.

Unsurprisingly, statements by groups of indigenous leaders go further. The SDGs do not challenge them, they challenge the SDGs. They claim that "our institutions uphold sustainable development ... We are in fact the embodiment of sustainable development" (Indigenous Peoples Major Groups 2015).

> To survive climate change ... we must heal the sacred in ourselves and include the sacredness of all life in our discussions, decisions and actions ... away from the commodification of Mother Earth. Neither world leaders nor modern institutions have the tools to adequately address climate change.
> (Indigenous Elders 2015)

Evangelical Christians, on the other hand, are conflicted. White American evangelicals are negative outliers on climate change (Guth 2010, 46). Nevertheless, an entity called the Micah Challenge (a global coalition of Christians holding governments to account for their promise to halve extreme poverty by 2015, named after a biblical prophet of social justice) was launched in 2004 to advocate globally among evangelicals for the MDGs; and now Micah Global is doing the same for the SDGs, producing material to help pastors preach in ways that make the SDGs applicable to the average person in the pews (ISJC 2017).

However, the lengthiest and most influential religious contribution to debate on the SDGs has come from a different source.

Laudato Si': religion questioning the sustainability of the SDGs

The Catholic Church, using its now lengthy tradition of social encyclicals, has responded to the SDGs (and the global reality which has led to the SDGs) with a book-length document, Laudato Si' (Pope Francis 2015). Several authors have contrasted Laudato Si' with the SDGs, and not in the latter's favour. Jason Hickel, for example, says Laudato Si' caused a stir around the world, whereas almost no one is excited about the SDGs. The encyclical is "visionary ... bold, uncompromising and radical"; the SDGs are "staid, timid and mired in a business-as-usual mentality". Hickel takes issue with three areas in particular. First, the SDGs are sprawling because they have confused thoroughness with holism. Pope Francis, by contrast, has struck at the systemic nature of the issue: "to seek only a technical remedy to each environmental problem which comes up is to separate what is in reality interconnected". Second, the SDGs are, according to Hickel, "a paean to consumption-driven economic growth"; the Pope, on the other hand, sees endless extraction as not just a physical impossibility but ultimately self-defeating and immoral. Fixes like carbon trading and renewable energy will not solve the problem. Third, the SDGs frame poverty and inequality as things that just exist, but the Pope dares to cast blame, calling out the transnational corporations, the foreign debt system and the over-powerful financial sector (Hickel 2015).

What Hickel's critique reminds us is that the papacy is relatively immune from the temptation to be a lowest common denominator. It can (sometimes) fulfil a prophetic role in relation to global sustainable development plans, thanks to three factors: first, its structural location on the margins of (but not totally outside) real political power and responsibility; second, the Catholic Church's straddling of global divides, giving it important constituencies in both the Global North and Global South; and third, its ancient tradition which pre-dates capitalism and whose scriptures are the most widely read pre-modern texts of all, coming to us like a voice from a strange world.

"I wish to address every person living on this planet … about our common home", says Pope Francis (2015, 6); and, by curious twists of history and our globalizing world, if anybody can, it probably is the occupant of the papacy. As monarchical head of the most global religious organization in the world, the Pope has "real power as convener, mediator and prophet in a world dominated by secular elites but peopled by resurgent religions" (Barbato and Joustra 2017, 2); the soft power capacity of the Pope "reaches beyond the Catholic flock into an emerging global public sphere" (3). The Holy See has observer status at the United Nations, enjoying a standing unique among religious bodies. It "represents [the] necessary but impossible task of integration" (3), being in tune with global ambitions which go far beyond the multiplicity of nation-states. The "solidarization" of international society, the trend to a world society of individuals with rights against states, has created a "window of opportunity" for actors such as the Pope (Diez 2017, 32), since it has given prominence to core values supposedly applicable to all of humanity, such as human rights, development, environmental concern and peace, as well as permitting the rise of para-diplomacy by non-state actors.

The Pope's encyclical on the environment is credited with exercising a significant influence on the Paris climate change talks of 2015 which occurred a few months later, being much discussed and explicitly mentioned by the leaders of several countries (Ware 2015). Nevertheless, Frecheville sees three areas in which the SDGs do not live up to the challenge of Laudato Si'. One is the economic critique of unlimited growth and, more broadly, of the "technological paradigm" and a "utilitarian mindset". Another is the political shortcoming of decision-making shaped by economic interests in favour of short-term profit. And last, the SDGs weakness on implementation, versus the Pope's criticism of countries which "place their national interests above the global common good", abandoning a culture of global solidarity for "the globalization of indifference" (Frecheville 2015).

In fact, the encyclical builds on the tradition of Catholic social teaching regarding economic models. Given the "alliance between the economy and technology" (Pope Francis 2015, 24), the environment is defenceless before "a deified market" (25). The verdict is categorical: "the environment is one of those goods that cannot be adequately safeguarded or promoted by market forces" (74). But Laudato Si' emphasizes links which go beyond the inadequacies of the market. "Our inability to think seriously about future generations is linked to our inability to ... [consider those] excluded from development" (64). The environmental crisis and the social crisis form a complex whole. And both are linked to a deeper cause. "Authentic human development has a moral character" (7). Our irresponsible behaviour has damaged both the natural and social environments; the "cry of the earth" and the "cry of the poor" are linked (22). What the encyclical calls the "dominant technocratic paradigm" (43) perpetuates "the lie that there is an infinite supply of the world's goods" (45) and "tends to absorb everything into its ironclad logic" (46) (a phrase reminiscent of Weber's depiction of the "iron cage" of bureaucratic capitalist rationality).

The ethical and spiritual roots of our crisis, says Laudato Si', require that we look for solutions not only in technology, leadership, new laws and even in a "true world political authority" (69), but also in a "change of humanity ... [through] an asceticism which ... is liberation from fear, greed and compulsion" (8). The majority of society must be personally transformed; "a commitment this lofty cannot be sustained ... without a spirituality" (83).

What sort of spirituality? Wilkinson (2015) notes that Pope Francis both refutes and confirms Lynn White's famous 1967 essay on the roots of the ecological crisis. White proposed Saint Francis as "patron saint for ecologists" but did so while referring to him as "clearly heretical" from the church's point of view. Pope Francis, says Wilkinson, rebuts that, taking a poetic line from his thirteenth-century namesake as the title of his encyclical, and suggesting that the heresy lies instead in the technologically magnified anthropocentrism which has commodified nearly all of creation.

But the encyclical has to walk a tightrope at this point, since the field of spiritually motivated critiques of anthropocentrism is already crowded. It thus takes aim not only at "a technocracy which sees no intrinsic value in lesser beings" but also at those (in some currents of biocentric spirituality) who, as the encyclical says, "see no special value in human beings" (Pope Francis 2015, 49) and who think "the presence of human beings on the planet should be reduced" (26). Demographic growth, Pope Francis insists, is fully compatible with an integral development. This leads on to another traditional Catholic emphasis: "concern for the protection of nature is ... incompatible with the justification of abortion" (50) which stems from the same "throwaway culture" (13) which generates waste. Another link is posited: "thinking we enjoy absolute power over our own bodies turns, often subtly, into thinking that we enjoy absolute power over creation" (62).

Thus, Laudato Si's "progressive" stances on economics and the environment come hand-in-hand with "reactionary" ones on gender, sexuality, and the family. Indeed, the document is relentless in its lack of inclusive language, highlighting the problem of an institution run largely by celibate males talking about the "encounter with someone who is different" (Pope Francis 2015: 62). It is true that the encyclical claims a necessary link between these "progressive" and "reactionary" positions, the same underlying logic. But, as Diez (2017, 32) says,

> the norms implicit in a solidarist international society provide a discursive context in which Christian norms ... related to peace, environmental sustainability and development have become easily translatable, whereas more personal norms on issues such as abortion or marriage have become more contested.

On the latter issues, the Pope becomes "merely one actor among many" (37).

None of this saves Pope Francis from the ire of the Catholic right. When the encyclical criticizes "the culture of relativism", for example, it is referring not to those who defend abortion, but to "those who say: Let us allow the invisible forces of the market to regulate the economy, and consider their impact on society and nature as collateral damage" (Pope Francis 2015, 50–51). A conservative Catholic author decries this as a "deeply negative view of free markets ... [reflecting] the conceptual apparatus of [1960s Latin American] dependency theory" (Gregg 2015). This critique of an unregulated economy, allied to the encyclical's argument that "technology based on the use of highly polluting fossil fuels ... needs to be progressively replaced without delay" (Pope Francis 2015, 65), has provoked the ire of American Catholic neo-conservatives (even if they often choose to attack the Pope on other grounds, e.g. Hickson [2015]: "has the Pope now [in welcoming the SDGs] given his moral support to an agenda ... contrary to the moral teaching of the Catholic Church on life and sexuality?").

Interestingly, Cafod UK (the Catholic Agency for Overseas Development) carried out a dialogue on Laudato Si' with its partner organizations from five global southern countries, and discovered some revealing differences between the encyclical and the perspectives of the participants. While the Pope was concerned with global ethical problems behind technical developments, the participants accentuated the advantages of technology for the poor. While the Pope saw political leaders as key drivers of change, participants were far more sceptical of their politicians. And last, the participants stressed gender equality as absolutely vital for sustainable development, an emphasis which was absent from Laudato Si' (Deneulin and Zampini-Davies 2017).

Conclusion: religion and the SDGs as mutual challenge and rival apocalypticisms

The literature on the SDGs provides examples from all the major religions of teachings and projects which support and seek to implement its ambitious agenda. All religions, it seems, can mobilize useful categories for "development work" and for "sustainability".

At the same time, many religions are sufficiently ambivalent in their teaching and ethics to provide support for divergent positions. An example of this is the division within evangelical Christianity, between unreconstructed climate change deniers and those who advocate urgently for climate change action. As Bergmann (2015, 395) says, we cannot make assumptions about the impact of religious ethics on how people think about climate-related questions.

As to the current position of FBOs, Marshall (2015, 386) concludes that "what seems to prevail today is an uneasy and often unclear set of relations…. Religious voices and beliefs are often influential in shaping many policies and ideas, but in a contested and often rather ambiguous manner", a situation which is compounded by the growth of religious illiteracy in many of the key donor countries.

Some arguments for the importance of religion are clearly overkill. If FBOs are so ubiquitous, credible, grassroots, and active, why is the world in its current plight? But one can, of course, always allege it would be worse otherwise; and the difficulty in divulging something like the SDGs (only 12 per cent of the population of Colombia have ever heard of them; and, more surprisingly, the same percentage of Danes [Risse 2017]) means that any extra help is welcome.

Meanwhile, one key country seems to be moving in the direction (albeit with a 1,000 qualifications) of granting religion a greater role: China (Laliberté 2015). Since the 1990s, religious entities (connected to the patriotic associations of the five official religions) have been able in practice to deliver some social services; but from 2012 they have enjoyed official support. There is evidence from fieldwork that an increasing number of people trust a variety of religious actors as agents of development. The Party, of course, does not yet invite religious leaders to express alternative views on development *policy*.

What of the future? Laudato Si' shows that some forms of religion might be the voices from "another world" that question, not the science but the presuppositions and the political forces behind the elaboration of the SDGs. It also underlines that religious communities can be especially influential in addressing what it calls the "cry of the earth" and the "cry of the poor", and putting both crises in integral and long-term perspective. This reminds us of Tocqueville's ([1835–1840] 2000) view of the role of religion in the "democratic era": that it would be needed, among other reasons, to promote non-material priorities and to play a counter-balancing role to democracy's innate tendency to short-termism. Given the inherent tensions in the idea of "sustainable development", perhaps religion's primary relationship to it can only be in terms of prophetic critique. The space for this, in the near future, will probably depend on Diez's depiction of the "window of opportunity" for such actors: will the "backlash against solidarization" (Diez 2017, 37) reflected in the rise of autocratic and populist tendencies narrow the window, or will a worsening of the environmental crisis open it even wider?

Thus, one of the challenges that religion places on development of any kind (the "new global morality, as proposed by the SDGs … is, in a way, the UN ecumenism" [Redo 2015]) is that of the mirror image. Beyond the dose of projection in the way development actors view religious actors, the fact is that the "older sister" is often further advanced in her self-criticism. Both humanitarianism and development evolved out of Christian missionizing, and both have been accused of perpetuating colonial attitudes and relationships. Yet the world would presumably be worse off without them (as Trumpism might yet show us). It is therefore unsatisfactory to think of the challenge of sustainable development without a sustained reflection on the rights and wrongs, and possible variations, of "proselytization" broadly conceived. The survival of the world, it seems, may depend on a successful "conversionist mission" – but exactly what is the message to be disseminated? If the Anthropocene can only be salvaged if anthropocentrism is overcome, which of the contrasting forms of anti-anthropocentrism is most suited?

Bergmann (2015) rightly stresses the reciprocal interaction between religion and environment. The various types of "greening" of religion, at least since the World Council of Churches coined the notion of "sustainable and just development" in 1974, have certainly not reached their limit. At the same time, other evolutions may kick in. Certain religious currents and the SDGs are, in a sense, rival apocalypticisms. Failure to meet the SDGs may well have catastrophic consequences which will have an immense impact on the religious field, reinforcing apocalyptic tendencies within Christianity and Islam as well as producing completely new religious movements. On the other hand, a new frugal puritanism might emerge, needing religious reinforcement, a Tocquevillean role for religion as the reconciler of democracy with a reality of limited economic possibilities and the need for long-term thinking. That may lead us back to Weber; after all, his key point was that such an epochal shift as the historic rise of a new economic system on the basis of bourgeois frugality needed a fundamental shift in values, supported by doctrinal justification and personal and communitarian motivation. Now, if the "iron cage" of capitalist rationality is leading to hell, a great force will be needed to free us from it.

Weber himself, in the early twentieth century, ruminated on such possibilities. "No one knows ... whether at the end of this tremendous development entirely new prophets will arise, or there will be a great rebirth of old ideas and ideals, or, if neither, mechanized petrification" ([1904–1905] 2005, 124). Perhaps, like him, we can speculate about three possible scenarios for our medium-term future. One possibility is that the environmental crisis will be more or less adequately resolved with technological adaptations involving minimal societal disruption, implying little change in the role of religion. A second scenario is that the crisis will only be overcome with considerable non-technological adaptations requiring substantial cultural change. In this case, one can imagine an enhanced and varied role for religion, as ideologue and also as promoter and (ascetic) motivator, as well as communitarian sustainer. Finally, a third scenario is that the crisis will not be adequately overcome at all, leading to catastrophic upheavals in which apocalyptic forms of religiosity will flourish, ranging from abstentionist forms of withdrawal from a presumed era of prophecy fulfilment, to radical new movements of "survivalist" religion.

References

Ager, A., and J. Ager. 2016. "Sustainable Development and Religion: Accommodating Diversity in a Post-Secular Age". *Review of Faith and International Affairs* 14 (3): 101–105.

Bäckström, A., and G. Davie. 2010. *Welfare and Religion in Twenty-First Century Europe: Volume 1: Configuring the Connections.* London: Routledge.

Barbato, M., and R. Joustra. 2017. "Introduction: Popes on the Rise". *Review of Faith and International Affairs* 15 (4): 1–5.

Basedau, M., Gobien, S., and S. Prediger. 2017. "The Ambivalent Role of Religion in Sustainable Development". German Institute of Global and Area Studies, GIGA Working Paper no. 297. Accessed 27 November 2018 www.giga-hamburg.de/en/system/files/publications/wp297_basedau-gobien-prediger.pdf.

Bergmann, S. 2015. "Sustainable Development, Climate Change and Religion". In *Routledge Handbook of Religions and Global Development*, edited by E. Tomalin, 389–404. London: Routledge.

Berkley Center. 2007. *Report of the Symposium on Faith-Inspired Organizations and Global Development Policy.* Washington: Georgetown University. Accessed 27 November 2018. https://berkleycenter.georgetown.edu/publications/176.

Beyer, P. 2006. *Religions in Global Society.* London: Routledge.

Bompani, B. 2015. "Religion and Development in Sub-Saharan Africa: An Overview". In *Routledge Handbook of Religions and Global Development*, edited by E. Tomalin, 101–113. London: Routledge.

Buddhist Declaration on Climate Change. 2015. *The Time to Act is Now: A Buddhist Declaration on Climate Change.* 14 May. Accessed 27 November 2018. http://fore.yale.edu/files/Buddhist_Climate_Change_Statement_5-14-15.pdf.

Clarke, G., and M. Jennings. 2007. "Introduction". In *Development, Civil Society and Faith-Based Organizations*, edited by G. Clarke and M. Jennings, 1–16. Basingstoke, Palgrave Macmillan.

Deacon, G., and E. Tomalin. 2015. "A History of Faith-Based Aid and Development". In *Routledge Handbook of Religions and Global Development*, edited by E. Tomalin, 68–79. London: Routledge.

Deneulin, S., and M. Bano. 2009. *Religion in Development: Rewriting the Secular Script.* London: Zed Books.

Deneulin, S., and A. Zampini-Davies. 2017. "How the Sustainable Development Goals (SDGs) Can Engage with Religion". Cafod, 19 October. Accessed 27 November 2018. http://eprints.lse.ac.uk/76459/1/How%20the%20Sustainable%20Development%20Goals%20%28SDGs%29%20can%20engage%20with%20religion%20_%20Religion%20and%20the%20Public%20Sphere.pdf.

Diez, T. 2017. "Diplomacy, Papacy, and the Transformation of International Society". *Review of Faith and International Affairs* 15 (4): 31–38.

Dijk, R. van. 2012. "Pentecostalism and Post-Development: Exploring Religion as a Developmental Ideology in Ghanaian Migrant Communities". In *Pentecostalism and Development*, edited by Dena Freeman, 87–108. Basingstoke: Palgrave Macmillan.

Duff, J., M. Battcock, A. Karam, and A. R. Taylor. 2016. "High-level Collaboration Between the Public Sector and Religious and Faith-based Organizations: Fad or Trend?". *Review of Faith and International Affairs* 14 (3): 95–100.

Erasmus. 2015. "Faith and Secular Global Bodies Learn to Live Together". *The Economist*, 19 September. Accessed 27 November 2018. www.economist.com/erasmus/2015/09/19/faith-and-secular-global-bodies-learn-to-live-together.

Fountain, P. 2013. "The Myth of Religious NGOs: Development Studies and the Return of Religion". In *International Development Policy: Religion and Development*, edited by G. Carbonnier, 9–30. Basingstoke: Palgrave Macmillan.

Fountain, P. 2015. "Proselytizing Development". In *Routledge Handbook of Religions and Global Development*, edited by E. Tomalin, 80–97. London: Routledge.

Frecheville, N. 2015. "The Sustainable Development Goals and Laudato Si". *Cafod Policy*. Accessed 27 November 2018. https://cafodpolicy.wordpress.com/2015/09/17/the-sustainable-development-goals-and-laudato-si/.

Freeman, D. 2012. "The Pentecostal Ethic and the Spirit of Development". In *Pentecostalism and Development*, edited by Dena Freeman, 1–38. Basingstoke: Palgrave Macmillan.

Gregg, S. 2015. "Laudato Si: Well Intentioned, Economically Flawed". *The American Spectator*, 19 June. Accessed 28 November 2018. https://spectator.org/63160_laudato-si-well-intentioned-economically-flawed/.

Guth, J. 2010. "Economic Globalization: The View from the Pews". *Review of Faith and International Affairs* 8 (4): 43–48.

Hefferan, T. 2015. "Researching Religions and Development". In *Routledge Handbook of Religions and Global Development*, edited by E. Tomalin, 36–52. London: Routledge.

Hickel, J. 2015. "The Pope v. the UN: Who Will Save the World First?" *Guardian*, June 23. Accessed 27 November 2018. www.theguardian.com/global-development-professionals-network/2015/jun/23/the-pope-united-nations-encyclical-sdgs.

Hickson, M. 2015. "The Pope and the UN's 'Development Goals'". *Catholicism.Org*, 4 November. Accessed 27 November 2018. http://catholicism.org/the-pope-and-the-uns-development-goals.html.

Hindu Declaration on Climate Change. 2015. "Bhumi Devi Ki Jai! A Hindu Declaration on Climate Change". 23 November. Accessed 27 November 2018. www.hinduclimatedeclaration2015.org/english.

Holman, S. R. 2015. *Beholden: Religion, Global Health, and Human Rights*. Oxford: Oxford University Press.

ICRC. n/d. *Code of Conduct for the International Red Cross and Red Crescent Movement and Non-Governmental Organizations (NGOs) in Disaster Relief*. Accessed 1 December 2018. www.icrc.org/en/doc/assets/files/publications/icrc-002-1067.pdf.

Indigenous Elders. 2015. "Indigenous Elders and Medicine Peoples Council Statement United Nations Convention on Climate Change, COP21 Paris, France". 30 November. Accessed 27 November 2018. http://nativenewsonline.net/currents/indigenous-peoples-release-joint-statement-to-un-talks-in-paris-on-climate-change/.

Indigenous Peoples Major Groups. 2015. "Statement of the Indigenous Peoples Major Groups to the United Nations Summit for the Adoption of the 2030 Agenda for Sustainable Development". 25–27 September. Accessed 27 November 2018. https://sustainabledevelopment.un.org/content/documents/19411indigenous-peoples-art.pdf.

ISJC (International Social Justice Commission). 2017. "Faith-Based Partnerships: Vehicles for Achieving the SDGs". *International Social Justice Commission, The Salvation Army*. Accessed 27 November 2018. www.salvationarmy.org/isjc/12july2017.

Islamic Leaders Climate Change Declaration. 2015. "Islamic Leaders Climate Change Declaration". *Ifees: Ecoislam*. Accessed 27 November 2018. http://islamicclimatedeclaration.org/.

Karam, A. 2015. "Religion and the SDGs: The "New Normal" and Calls for Action". *Inter Press Service*. Accessed 27 November 2018. www.ipsnews.net/2015/07/opinion-religion-and-the-sdgs-the-new-normal-and-calls-for-action/.

Laliberté, A. 2015. "Religion and Development in China". In *Routledge Handbook of Religions and Global Development*, edited by E. Tomalin, 233–249. London: Routledge.

Marshall, K. 2011. "Development and Faith Institutions: Gulfs and Bridges". In *Religion and Development*, edited by G. ter Haar, 27–53. London: Hurst & Co.

Marshall, K. 2015. "Complex Global Institutions: Religious Engagement in Development". In *Routledge Handbook of Religions and Global Development*, edited by E. Tomalin, 373–388. London: Routledge.

Noor, Z., and F. Pickup. 2017. "Zakat Requires Muslims to Donate 2.5% of Their Wealth: Could This End Poverty?" *Guardian*, 22 June. Accessed 27 November 2018. www.theguardian.com/global-development-professionals-network/2017/jun/22/zakat-requires-muslims-to-donate-25-of-their-wealth-could-this-end-poverty.

Occhipinti, L. 2015. "Faith-Based Organizations and Development". In *Routledge Handbook of Religions and Global Development*, edited by E. Tomalin, 331–345. London: Routledge.

Olivier, J. 2016. "Hoist by Our Own Petard: Backing Slowly Out of Religion and Development Advocacy". *HTS Teologiese Studies/Theological Studies* 72 (4), a3564. Accessed 27 December 2018. https://jliflc.com/resources/hoist-petard-backing-slowly-religion-development-advocacy/.

Pew Forum. 2012. *The Global Religious Landscape*. Accessed 27 November 2018. www.pewforum.org/2012/12/18/global-religious-landscape-exec/.

Pope Francis. 2015. *Laudato Si'*. Vatican: Libreria Editrice Vaticana.

Redo, S. 2015. "Religion and the Sustainable Development Goals". *Universal Peace Federation*, 6 February. Accessed 27 November 2018. www.upf.org/resources/speeches-and-articles/6644-s-redo-religion-and-the-sustainable-development-goals.

Risse, N. 2017. "Getting Up to Speed to Implement the SDGs: Facing the Challenges". *IISD*, 25 April. Accessed 27 November 2018. http://sdg.iisd.org/commentary/policy-briefs/getting-up-to-speed-to-implement-the-sdgs-facing-the-challenges/.

Ryan, J. 2012. "Washing our Dirty Feet: Christ and Development". In *Mission and Development*, edited by M. Clarke, 17–28. London: Continuum.

Santis, S. de. 2016. "Faith Groups Essential to Achieving Development Goals". *WACC*, 4 April. Accessed 27 November 2018. www.waccglobal.org/articles/faith-groups-essential-to-achieving-development-goals.

Selinger, L. 2004. "The Forgotten Factor: The Uneasy Relationship between Religion and Development". *Social Compass* 51 (4): 523–543.

Seneviratne, K. 2017. "Eastern Spirituality Could Help Sustainable Development". *IDN-News*. https://archive.indepthnews.net/index.php/the-world/asia-pacific/1305-eastern-spirituality-could-help-sustainable-development.

Sidibé, M. 2016. "Introduction: Religion and Sustainable Development". *Review of Faith and International Affairs* 14 (3): 1–4.

Thomas, S. 2005. *The Global Resurgence of Religion and the Transformation of International Relations*. New York: Palgrave Macmillan.

Tocqueville, A. de. (1835–1840) 2000. *Democracy in America*. New York: Bantam Classics.

Tomalin, E. 2015. "Introduction". In *Routledge Handbook of Religions and Global Development*, edited by E. Tomalin, 1–13. London: Routledge.

UN Chronicle. 2008. "Tackling Poverty Reduction: The Role of the Islamic Development Bank". *UN Chronicle* XLV (1), March. Accessed 27 November 2018. https://unchronicle.un.org/article/tackling-poverty-reduction-role-islamic-development-bank.

Ware, J. 2015. "COP21: Laudato Si a Major Talking Point at Climate Change Talks in Paris". *The Tablet*, 6 December. Accessed 27 November 2018. www.thetablet.co.uk/news/2885/cop21-laudato-si-a-major-talking-point-at-climate-change-talks-in-paris.

Weber, M. (1904–1905) 2005. *The Protestant Ethic and the Spirit of Capitalism*. London: Routledge.

Wilkinson, L. 2015. "Did Pope Francis Study at Regent?". *Regent World* 27 (2), Fall. Accessed 27 November 2018. https://world.regent-college.edu/leading-ideas/did-pope-francis-study-at-regent.

11 The ecological limits of the Sustainable Development Goals

Stephen Quilley and Kaitlin Kish

Introduction

The updated Sustainable Development Goals (SDGs), which include ending poverty, promoting universal education, gender equality, health, environmental sustainability, and global partnerships, all depend to a significant degree upon a high and continuing throughput of energy and materials. In elaborating its SDGs, the UN begins the definition of the overall goal for the Anthropocene with "development": "Development that meets the needs of the present while safeguarding Earth's life-support system, on which the welfare of current and future generations depends" (Griggs *et al.* 2013, 305). The policy vehicles required to meet such a goal necessarily rely on ongoing fiscal transfers generated by ongoing development and a growing economy. Consider for example: institutional education; the innovation, manufacture and delivery of vaccines; governance systems to advance the position of women; the delivery of multiple systems of cheap contraception; market and state pension systems that relieve aging citizens of dependence on uncertain family arrangements. In different ways, all of these policy commitments involve the replacement of less complex systems that once existed in the informal sector, with more complex systems in the domain of the formal economy, the state, and the market. Moreover, one of the SDGs is explicitly to promote economic growth as measured by GDP, labour productivity, and access to financial services.

Ecological economics very clearly demonstrates the tension between biospheric limits and current trajectories in economic growth. The scale of material and energetic flows of the economy relative to those of ecological systems are firmly out of balance (Daly and Farley 2011). Thus, from any serious ecological perspective, predicating sustainability goals on a dependence of development on growth is very problematic. In a recent paper, O'Neill *et al.* (2018) quantified the resource use associated with meeting basic human needs for over 150 countries. Comparing this to what is deemed to be globally sustainable, they found that there is currently no country that achieves a good life for its citizens at a level of resource use that could be extended to all people on the planet. This suggests a prima facie case for the proposition that the SDGs are intrinsically flawed and that they cannot in principle deliver "sustainable development". In what follows we develop the case for Ecological-Economic Goals (EEGs) that have a better chance of reconciling development with the integrity of the ecological life support systems of the biosphere.

The failure of sustainable development is a function of the failure of "ecological modernization" strategies to deliver the long promised "dematerialization" and the "decoupling" of growth from resource use and waste – as alluded to in Blay-Palmer and Young's chapter; there are many similarities between agroecology and ecological economics. In economic terms this reflects the persistent operation of the Jevons paradox, i.e. the tendency for efficiency gains to fund renewed investment, technical innovation, more rapid product cycles, and further expansion in the scale of production and consumption. In social and political terms, such failure relates to the systemic resilience of consumerism as a source of jobs, fiscal revenues, and individual and collective security; and, culturally, to the hegemonic and ubiquitous role of consumption as a source of meaning and ontological security (Dickinson 2009).

In the face of such paradigmatic failure, any feasible environmental politics must be rooted in:

1   A non-negotiable recognition of biophysical limits and the priority that must be accorded to the scale of economy relative to ecology – from the ecological economic tradition associated with Herman Daly, Georgescue-Rogen, H. T. Odum, Charles Hall, Robert Constanza and others.
2   A historical sociological recognition of the historicity of psychology/personality, institutions and ideas – from the historical sociological and anthropological tradition associated with Norbert Elias and Karl Polanyi.
3   A deep recognition that any paradigm shift regarding the metabolism of society will involve equally new, distinctive, and possibly uncomfortable (combinations of) ideas, institutions, perceptions, and even (in the long run) personality types.

In this chapter we briefly explore these three theoretical foundations, focusing on the discourse of biophysical limits and pervasive processes of modernity. We use this as a backdrop to argue for EEGs that move beyond the paradigm of growth, into post-growth goals for change.

Limits thinking and Enlightenment universalism

Starting from the explicit premise that there are limits to growth, our argument uses Ecological Economics, the only framework that starts from an understanding of both the economy and society as embedded within and a function of ecological systems. For Daly this relationship could be summarized by the triptych subordination of efficiency/markets to social justice, which in turn was conceived as subordinate to the ecological/metabolic scale. Ecological economists attempt to develop solutions that respect this hierarchical system, which has been largely ignored by mainstream economics, as well as national strategies, policymakers and politicians, and even environmentalists devoted to the idea of sustainable development or green growth.

Many social ailments explored by sociologists and social theorists, such as alienation, narcissism and ontological security, are a direct, and potentially unavoidable, cost of capitalist modernization and of the process of individualization, which is intertwined with the development paradigm (Weber 1968; Ollman 1977; Giddens 1991; Durkheim 2006; Twenge 2017). They cannot necessarily be "solved". At best, society can make choices about a balance of "goods and bads". The disembedding of traditional economy and society (Polanyi 1994; Quilley 2012) was traumatic and violent. In early Modern England, both "free-wage labour" and the modern citizenry alike were made possible by the enclosure of the commons and the expropriation of the peasants. But this very doubled-edged process of "emancipation" also engendered a process of individualization – without which it is hard to imagine positive phenomena such as extended empathy (Rifkin 2009), scientific reasoning, multiculturalism, increased rights for many (Pinker 2012), and the modern idea of the sovereign individuals necessary for democracy and rights-based politics (Elias [1987] 2010; Quilley 2012, 2017).

That the UN's Sustainable Development Goals are unlikely to work should be no surprise. They are very explicitly premised on generalizing a pattern of life that emerged initially early modern England and became generalized in the universalist vision of Kant, Rousseau, Condorcet and other Enlightenment philosophers. The original sin of liberal society is that "freedom" was in large part imposed. Its discomforting secret is that this kind of freedom, though sacred and foundational for all other cherished dimensions of modern liberal society, comes with a high price tag.

With respect to the SDGs, the most significant aspect of this cost relates to what Odum (2007) refers to as the "transformity cost" of social complexity. In Polanyi's terms, the emergence of disembedded, price-setting markets was, from the outset, profoundly destabilizing. Undermining social cohesion and leaving millions of free wage labourers at the mercy of the market, the liberal-market vision was, he argued, a utopian project. Left to itself, the utopian project of the Market Society would destroy both nature and the social fabric. For this reason, argued Polanyi, market liberalism almost immediately engendered a "countervailing movement for societal protection". This manifested itself in the gradual extension of the regulatory state (what Bourdieu has called "the left hand" of the state [Loyal and Quilley 2017]) over all aspects of social reproduction that were formerly the exclusive province of families, communities, feudal estates, or religious institutions.

In this way, capitalist modernization involves above all the disembedding of individuals, the unravelling of local place and kin-based forms of reciprocation, and the emergence of the formal economy as a distinct sphere that is intimately tied to the emerging nation state. The process of "disembedding" involves the relentless expansion of a transactional economy involving increasingly instrumental-rational individuals in the context of mutually dependent institutions of market and state. Elias (1987) construes this process as a shift in the scale and scope of the "survival unit" from family/place-bound community to that of individual-citizen/state/market. But significantly, dominant "common sense" conceptions of progress, freedom, and affluence have also come to be understood through the same lens, i.e. the capacity of individuals to pursue their interests and goals, to make choices, to behave, to move around, to make/unmake relationships unconstrained by any ascriptive social context.

As has already been intimated above, in the early modern period there were many bottom-up, communitarian social innovations (Friendly Societies, Guilds etc.) that sought to create symbolic forms of familial reciprocation and security (Quilley 2012). These communitarian experiments would be in the bottom right-hand corner of Figure 11.1. These can be seen as transitional attempts to

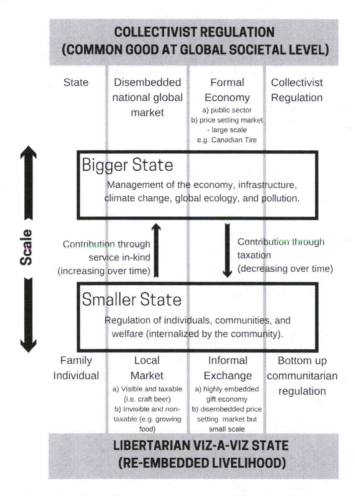

Figure 11.1 The politics of state, market, and livelihood.

Source: created by authors.

shield increasingly mobile individuals from the vagaries of the market. But in the event, from the end of the nineteenth century such social innovations were displaced by expanding state institutions. And in the twentieth century, social progress has been conceived in a very consistent way as a function of the State balancing the Market – a dyad that mapped neatly onto a binary left–right political spectrum. Political debates have focused on the balance of state and market and in more recent decades on the extent to which the state uses market-like mechanisms to allocate goods and services. Nevertheless, the elimination of bottom-up forms of communitarian social insurance rooted in the domain of what Polanyi called "Livelihood" has been an invisible default mode for both public policy and debate.

How might we understand this process through the lens of ecological economics? In essence, more complex institutions and disembedded processes, involving a greater throughput of energy and materials, more transactions in the formal economy, and a higher "transformity" value, have consistently replaced lower-complexity and more embedded forms of provision associated with the domain of livelihood. Thus, for example, informal, familial care of people with disability has been supplemented or even displaced by special schools, residential care units, complex processes of legal regulation, and a labyrinthine architecture of welfare benefits directed at individuals. The sector is predicated on a highly disembedded and visible array of transactions and complex system for training and regulating the educators. The same kind of shift from (embedded) household/livelihood systems of support to much more complex, disembedded, individually focused systems – can be seen in every arena of societal reproduction.

Together, this expanded tissue of state activities as a backdrop to a society of mobile, instrumental-rational, and sovereign individuals, disembedded from the ascriptive fabric of traditional society, has become the taken-for-granted landscape of social reality – not just a description but a not-so-subtle prescription. And because the SDGs were created unconsciously with this reality and prescription in mind, they reflect and valorise the idea that social progress must be a function of expanding social complexity, an elaborating mesh of state activities, and more extensive markets, all of which presuppose growth.

Our position is that, as presently conceived, the SDGs are wedded to this paradigm of modern individualism, and as such are unable to produce the kind of change necessary for real socio-ecological change. Instead, we seek to advance genuinely sustainable **EEGs** – goals which necessarily involve the re-emergence of livelihood and community as one principle of social and economic organization, and which balance both the state and market (Kish 2018).

Sustainable Development Goals, complexity and growth

No one will argue that the SDGs are inherently bad. Who could argue with the eradication of poverty, hunger, and ill-health? The problem is that the common sense understanding of each goal is firmly rooted in a modern and individualistic frame of reference. If "no poverty" means every country following the Western trajectory of industrial modernization, then we have a very clear ecological conflict with the goal. This kind of conflict is replicated in relation to every specific goal. Moreover, what is significant about these examples is not necessarily the headline goal or target (e.g. security in old age), but that the implicit range of solutions reflects deep ontological assumptions about the nature of modernity. Some key assumptions are:

1   Social justice and security aspirations must refer primarily or only to individual citizens.
2   The agents involved are spatially and socially mobile, rational individuals.
3   Security is primarily a function of the relation between individual and state.
4   Progress will likely see the further consolidation of this survival unit at the expense of traditional (clan, family, place, community, estate) forms of security and mutual identification.
5   The number of separate households per capita is likely to rise as a function of the dissolution of extended families, reduced social pressures against divorce, individual living patterns, etc.

For example, consider SDG 3 (Good Health and Well-Being). UN indicators refer variously to the need to: reduce maternal and infant mortality ratios to "Western norms"; end epidemic diseases; greatly extend prevention and treatment for narcotic abuse and addiction; extend systems to reduce road accident deaths; ensure universal access to sexual and reproductive health systems and achieve universal health coverage; extend state and corporate governance systems to reduce death and illness from chemical pollution; strengthen the capacity of developing countries, for early warning, risk reduction, and management of national and global health risks.

Other things being equal, these are all laudable aims which, taken together, represent an aspiration to generalize Western standards and approaches to health and well-being. But of course, at the risk of repetition, such systems are "entropically expensive". Without exception, they involve the supplanting of less complex, more embedded forms of care and management largely in the domain of families and communities, with more complex forms organized by the state (represented by a shift from the bottom to the top in Figure 11.1). It is easy to see that such systems are more expensive in financial terms. Because they elaborate complex organizations with paid labour forces, training and standards protocols etc. they depend entirely on fiscal transfers from a growing economy. The near collapse of the health system in post-Soviet Russia and Cuba (in the 1990s) and again in Greece in 2008–2010 provide visceral examples of this dependence. But following Odum (2007), it is also clear that such systems involve the elaboration of objects and processes that involve more energy transformations (with higher transformity values) further along the energy hierarchy. Take, for example, care by a professional nurse as opposed to a child's mother. The nurse has to be trained by health educators, in a teaching hospital, which has to be provisioned, heated, regulated, inspected, audited, periodically upgraded. S/he then has to be paid on an ongoing basis. Tracing out even such a simple system involves an almost innumerable series of financial transactions and energy transformations – which ultimately involve a very direct ecological (carbon, water, materials, pollution) footprint.

At the same time, the implied programme of health care seems also to embody the same ontological and sociological assumptions (A–E) referred to above. But the pattern of life associated with this version of modernity is associated with its own set of health problems which incur further energetic and ecological costs. For instance, a large contributing factor to the decline of mental health in Western society is this excessive differentiation of the individual and the self-perception of separation, autonomy, and independence (Twenge and Foster 2008; Donnellan, Trzesniewski, and Robins 2009; Twenge and Campbell 2010; Waugaman 2011). Twenge and Foster argue that this obsession is not an organic human development, that it has been engineered by advertising companies, social media, and consumer society, which creates a feedback loop. More realistically it should be understood as a function of modernity, i.e. complexity and the extension of the social division of labour; the process of disembedding; the loss of connection with place-bound communities; and individual relations to market and state (Beck 1992; Elias [1989] 2007). Nonetheless, this perception of the self as sovereign and self-determining is sold to modern individuals through advertisements and reinforced through the labour market and through mechanisms of social and private insurance (see Quilley 2019).

Taking into account SDG 4 (Quality Education), indicators once again embody rather explicitly the same ontological assumptions (A–E). The goals refer to universal numeracy and literacy, and gender-blind, free, and equitable access to high quality primary and secondary education systems with explicit reference to the "gender sensitive" inclusion of women and people with disability. Quite clearly, this goal reflects a commitment to the idea of rational individual agents who are able to effectively engage with the institutions of state, civil society, and market economy *as individuals* and in ways only minimally mediated by lower level group affiliations.

In this way, this vision unconsciously reflects the priorities of what Ernest Gellner (1983) referred to as "exo-education" i.e. the replacement of informal and local/familial processes of acculturation by monopolistic agencies of state. For Gellner the principle objective of such exo-education was to impose a shared "high culture" involving standardized norms, categories, and ontological-cognitive frameworks – and equipping individual citizens with the skills and competences necessary to engage effectively with the labour market and the institutions of the nation state. In the nation-state formation, education is the field which most deliberately articulates the process of individualization and the corollary transition from tribal/local "survival units" to that of the "imagined community" of the nation-state.

There is no doubt that such processes of exo-education are a prerequisite for the project of democratic state formation and the advancement of both citizenship-based entitlements and more general human rights. And since the sometimes more, sometimes less coercive elaboration of a shared national "high culture" marches in tandem with a whole set of wider institutional capacities, including the state's monopoly on violence, SDG 4 is intimately related to the achievement of other SDGs, notably SDG 8 (Decent Work and Economic Growth), SDG 9 (Industry, Innovation and Infrastructure), and SDG 16 (Peace, Justice, and Strong Institutions).

However, none of this is necessarily sustainable. For a start, the relationship between modernizing education systems designed to create able, individual citizens and polyvalent workers, both depends upon and is a driver of growth. Complex education systems depend directly on the regular flow of fiscal resources from a growing economy. At the same time, an effective education system provides the skilled workforce necessary for ongoing processes of technical innovation and economic transformation.

The imposition of a single language or high culture necessarily involves a winnowing of subaltern cultures, with thousands likely to disappear in coming decades. Ironically, this process of homogenization and rationalization upon which all the other SDGs depend must involve something tantamount to cultural genocide – at least in (most) states in which often hundreds of potential language-nations jostle for recognition. With regard to cultural genocide, Article 7 of the 1994 Draft of the UN Convention on the Rights of Indigenous Peoples includes reference to "Any form of assimilation or integration by other cultures or ways of life imposed on them by legislative, administrative or other measures" (UNWGIP 1994).

More generally, the process of state sponsored exo-education is part of a deliberate unravelling of traditional communitarian forms of society often rooted in shared religious cosmologies and in the rhythms and seasonality of rural societies structured around common pool resources (Ostrom 1990). This orchestrated deconstruction of the rural gemeinschaft accelerates the Weberian processes of individualization, rationalization, and secularization that are unfolding in the great movement of peasants into the cities.

One of the most challenging goals relates to gender equality and sensitivity – SDG 5 (Gender Equality). The indicators for SDG 5 refer to ending discrimination, eliminating violence and sexual exploitation, and tackling the problem of unpaid household/caring labour through the provision of services (elder care, childcare, disability services etc). As well as promoting equality within the home and the extension of reproductive and sexual health services, the provisions also refer to equal participation and leadership in all areas of economic, cultural, and political life.

Whatever else, this SDG is very explicitly directed at the unravelling of traditional forms of rural society which are predicated upon ascriptive social roles and a highly gendered division of labour. As we have seen, the involuntary and forcible disembedding of individuals from traditional roles and obligations is a defining feature of modernization. Certainly, this can be construed as a project synonymous with the Enlightenment vision of social emancipation. But as with the domains of health and education, it is also deeply implicated in the process of state formation and the creation of the society of individuals. The Market Society is also necessarily a Civic Society of Sovereign Citizens.

But, once again, even discounting the possibly Eurocentric and even neo-colonialist overtones of this universalist understanding of individual rights, an unavoidable corollary of emancipatory gender-individualism is the growth of complexity. Many, if not all, of the specific policy vehicles designed to achieve SDG 5 rely firmly on processes deeply dependent on economic growth: e.g. institutional education; affirmative action; childcare; complex gender-specific health provisions; contraception; state (income, care, housing) support for single parents; and state safety nets designed to work alongside the policy of instant, fault-free divorce. In each case, relatively simple, embedded, and energetically cheap arrangements in the informal/gift economy are replaced by administratively complex, multi-tiered, financially expensive, transactional, and regulated systems in the formal economy, tied either to the state or the market or both. It follows that any process of "degrowth", economic contraction, or relocalization in the wake of energy/resource constraints would necessarily see a shift in the opposite section (Figure 11.1). Having given women enormous freedom over their bodies and lives and a driver of the feminist vision of emancipation through the labour market, birth control provides a salutary example of the tension between social and environmental priorities (Heer and Grossbard-Shechtman 1981; Katz and Goldin 2002). An IUD requires the creation of plastic and mining of copper, and pills require the high tech, biochemical synthesis and production – necessitating complex systems and an economic division of labour involving hundreds of millions of people. And as with the other examples, following Odum (2007), the associated energy footprint is not just embodied in the particular artefact or processes of development, but distributed across an incredibly complex networked hierarchy of energy transformations (Zywert and Quilley 2017). Exactly the same considerations apply to childcare arrangements. From the 1960s feminist advocacy focused on emancipation through the labour market and demands for the state to take on responsibility for both child and elder-care, equalizing the position of men and women as citizens-consumers-workers. In effect this involved shifting the care functions from the (now residual) pole of livelihood towards systems of state and market (from the bottom to the top of Figure 11.1).

Summary

The deep, egregious tension between sustainability and development is perhaps captured most obviously by SDG 12 (Responsible Consumption and Production), which is associated with just two indicators of progress (UNGA 2015, 21–22):

- "Decoupling economic growth from natural resource use is fundamental to sustainable development".

- "Countries continue to address challenges linked to air, soil and water pollution and exposure to toxic chemicals under the auspices of multilateral environmental agreements".

This is egregious because it seems wilfully blind to 50 years of failure in relation to the technical project of ecological modernization (compare Rockström *et al.* [2009] with Mol and Spaargaren [2000]). Similarly, two goals, SDG 8 (Decent Work and Economic Growth) and SDG 9 (Industry, Innovation, and Infrastructure) are in direct and obvious conflict with any low-growth agendas. However, by the same token, these goals are prerequisites for the SDGs discussed above that relate to a broad swathe of social justice targets. That is so precisely because the latter are predicated on a shift from less complex systems rooted in the domain of family/community/livelihood to more complex systems orchestrated through state, market, and directed at citizens, consumers and rational individuals (see Figure 11.1).

State, market and livelihood

If sustainable development founders in part because of a pre-analytical commitment to the society of individuals (see Quilley 2019), then what might be the alternative? What shape and institutional architecture would it have? What kind of binding culture and ontology? And how would these features shape the "average personality" structure of citizens? How might the I/We balance begin to change (Elias [1987] 2010)? What might a feminist agenda look like in such a context?

One possibility suggested by Polanyi's account of modernization would involve the partial re-embedding of the market economy. In what follows, we suggest that the alternative to a one-dimensional reliance on the nation-state would involve a different kind of "re-embedding" – namely the re-emergence of the familial, community, locality-based principle of *livelihood* as a triangulating counterpoint to both the *state* and the *market*. Such re-embedding through livelihood shares an affinity with the notion of *subsidiarity* that underpins the social-catholic vision of Distributism (Médaille 2010). Developed by G. K Chesterton as an alternative to both capitalism and socialism, this putative "third way" intimates the ways in which the re-emergence of livelihood might unveil political scenarios in the adjacent possible that are rendered invisible by the left/right polarity of state versus market.

Livelihood in Polanyi's sense retains the possibility of a lower transformity/complexity version of modernity in Odum's sense. This is because any such transition would involve a shift from more elaborate and institutionally differentiated state/market systems geared to the life course of free-floating individual citizen-consumers, to less complex, less specialized forms of provision rooted in hearth or community and oriented to figurations of interdependent individuals tied together by processes of structured and culturally endorsed reciprocity.

Partially reversing many of the extensions of state and market mechanisms that have proliferated since the nineteenth century, a political economy of livelihood would reduce the number and scale of energy transformations and "transformity values" associated with particular functions. Thus, for instance, informal reciprocal childcare arranged between neighbours or siblings involves much less embodied/distributed energy than the elaborate institutionalized forms that have expanded over the last 40 years.

Quite clearly any significant processes of relocalization and re-embedding would involve a systemic shedding of some of this complexity along with the contraction of the economy and the informalization of activities that for the last two centuries have been moving from the domain of livelihood into the quantifiable, fungible domain of the formal economy – orchestrated by companies and visible to the state (Scott 1998; Quilley 2015). The recovery of livelihood as a counterpoint to both state and market would point also to the re-emergence of early-modern patterns of solidarity and mutualism in the advanced economies of the Global North (and the retention and repurposing of pre-modern, agrarian principles of solidarity in the Global South (Zywert and Quilley 2019).

In practical terms any movement in this direction will always be hampered by an underlying ubiquitous commitment to the rationality of actors as sovereign individual-citizens. Elias's ([1987] 2010) work demonstrated an intrinsic connection between culture and "average personality", and even to average patterns of perception and cognition. Over the last century, the competing political economies of left and right have taken for granted this landscape – with the focus for contestation being the boundary between the market and the state. A premise of the putative alternative "basin of attraction" is that the re-emergence of the domain of livelihood – self-sufficient provisioning, maintaining of body and soul in the context of extended family, community, and on the basis of gifting, reciprocity, and mutual aid – might open new ways for security, welfare, and livelihood to be achieved without the continual expansion of the state-market society. Such a project resonates with green visions of bioregionalism, anarchist and libertarian visions of the state-less society, virtue ethics and the social-Catholic project of distributism (e.g. Milbank and Pabst 2016; Dreher 2017; Médaille 2010). Taking Polanyi as a point of departure, our framework describes a region of the "state space" as a flexible and negotiable balance between state, market, and livelihood. We now go on to consider the kind of indicators – EEGs – that might be congruent with such a political economy of livelihood.

Alternative Ecological Economic Goals (EEGs)

Synthesizing insights from literature on historical sociology (Giddens 1991; Beck 1992; Bauman 2000) and case studies of alternative economies (Quilley 2012; Bauhardt 2014; Kish, Hawreliak, and Quilley 2016; Zywert and Quilley 2017, 2019) we articulate a new, community-based approach the SDGs.

Table 11.1 defines each of these EEGs and states' associated challenges and opportunities. The "challenges" relate to potential nonlinear and/or unintended consequences that may be associated with the pursuit of an EEG (Kish and Quilley 2017). The "opportunities" are based on possible ecological economic applications that might facilitate the development of the EEG in question. The opportunities column of this table is not exhaustive and can be built upon significantly. The idea is that the EEGs would be used as benchmarks against which to explore how, and the extent to which, communities may already be meeting those goals; and the role of ecological economic tools and techniques in facilitating further progress.

These EEGs are certainly normative, but no more so than the SGDs. They stand out because they embody a set of principles around the triptych of livelihood, state, and market – principles that represent a stark contrast to those more implicit and unacknowledged predicates in the SDGs. The purpose is not to claim that the achievement of these eight goals would lead to sustainable relationships between the biosphere, society, and economy. Rather the case is made that, together, these EEGs represent a new way to approach community-based problem solving, as well as a frame for imagining an ecological-economic trajectory for development. If the UN SDGs are incompatible with biosphere integrity, we need to rethink some of our fundamental ideas of what "development" means. Using this framework, any future ecological economic tool or technique should contribute to at least one EEG and, as far as possible, not detract from the integrity of the others.

It is also true that in some measure the EEGs embody insights from the more radical tradition of developmental thinking that goes back to Tolstoy, Geddes, and Gandhi (Bakker 1990; Mantena 2012) and, just prior to the consolidation of the Sustainable Development paradigm, to Schumacher (1989) and the appropriate technology movement. Critics from the 1970s, such as Alexander (2005), Alexander, Neis, Anninous, and King (1987), Jacobs ([1978] 2011), Turner (1976) and Ward (1976) looked to the favelas and shanty towns of South America, identifying a generative process of self-organization that produced more organic, functionally coherent, and cheaper forms of community design and architecture than top-down development. Their insights chimed with the wider movement for "appropriate technology" (CHF-BRI 1983; Hazeltine and Bull 2003) that drew explicitly on Gandhi and Schumacher as well as the highly influential work of Illich (1973) and radical innovations such as China's "barefoot doctor" programme (Zhang, Kleinman, and Tu 2011). They have also been echoed in more recent work on the Squatopolises of the Global South by Neuwirth (2006). The unifying thread is an implicit commitment to finding forms of technology and linked social institutions that can deliver "development" and some vision of modernity at a much lower "price point" in terms of the energy and material flows, as well as the prerequisite fiscal transfers drawn (ultimately) from the market sector. All of these authors and theorists understood implicitly that the Western model of consumer society, extreme forms of individualization and a top-heavy state/market complex probably could

Table 11.1 Ecological Economic Goals

EEG	Challenges	Opportunities
Shared Ontology *Belief systems, rituals consolidating normative and relational lived worldview*	Extreme difficulty of recovering shared religious/spiritual frameworks in cosmopolitan/secular society	Innovation of shared public rituals to consolidate and to some extent enforce a process of deliberative (ecological/relational) conscience formation
Equity and fairness *Just distribution of goods, services, and financial resources*	Difficult to manage/govern *from above*; requires organized complexity; requires an extended empathic circle	Open source access; local currencies; imaginative finance structures; mandatory work/life balance – just distribution
Subsidiarity *Maximum freedom of action to subsist at level of hearth and community*	Libertarian freedoms of individuals and families (from liability, red-tape, health and safety) in tension with need to regulate corporations and larger entities	Opportunity for localist, DIY culture of self-actualization outside the formal economy Grown and consolidate bridging and bonding capital within and between networked communities
Strong mental health *High self-esteem and life satisfaction*	Tension between re-enchantment or ontological security and instrumental rationality; decreased consumption and contraction of formal economy means fewer taxes and social services	Open-source/local currencies shown to improve self-esteem; citizen owned production; maker/family-based hero-projects; work/life balance such as 4-day work week and $70k income cap
Gender equality *All community roles viewed equally; but acceptance of both equality of opportunity and validity of traditional gender roles*	Contracting state likely to reduce the availability of formal child-care; challenges much modern feminism in so far as predicated on access to the formal economy	Work/life balance – four-day work week, home-life seen as equally valuable/empowering as work life; community parenting; re-emergence of extended family

Systemic education *'Endo' education supplements formal 'exo-education': meets the needs of the community*	Need to limit fear around harm/risk; in formal education systems, experiential education is expensive; requires continuity in place and greater intergenerational continuity	Embeddedness in place as a driver for what's learned; open-source access to education; students taught how to produce goods; hands on-learning; greater role of mentors Re-emergence of family/community contexts for learning. Greater autonomy of schools from boards, and boards from state
Redefined success *Non-monetary views of success*	Difficult to convince and engage people unless alternative meaning frameworks are introduced simultaneously i.e. *requires cultural transition*; barter/trade/gifting entails more limited access to more limited range of goods	Work/life balance including $70k income cap; innovation for needs – efficient allocation; encourage barter, gift, and trade rather than monetary exchange
Thoughtful consumption *Purchase of special products, more local consumption*	Contraction of formal economy will reduce tax transfers and engender fiscal crisis; breakdown of international trade relationships	Citizen-owned production creating special and/or more durable, higher quality and certainly more expensive products; embeddedness in place; needs-based innovation; hold repair cafes and artisan shows
Connection to place *More meaningful relationship to land; re-embedding of individuals, communities and ecological places*	Limits social and spatial freedom and mobility. Danger of exclusive ethno-spatial identifications	Embeddedness in place through cultural ecological restoration practices; work/life balance as a driver to "local tourism"
Profound community orientation *Normative pressure for transactions in the economy meet the needs of the community*	Likelihood of increased out-group antagonism	Citizen owned production as a way to learn/interact; work/life balance; open-source as a method of trust building; embeddedness in place; limit social media

Source: created by authors.

not be sustained in the West, let alone transferred to the Global South. They sought to reconcile an indictment of aggressive consumer capitalism with a commitment to a different kind of modernity. For example, the Centre for Alternative Technology in Machynlleth, Wales, played an important role in pulling countercultural ideas and techniques (passive solar, photovoltaics, biogas, composting, domestic permaculture, self-build housing) often drawn explicitly from Third World experience and innovation into the mainstream of UK sustainability discourse and policy. There is then a body of experience and research to build upon.

Implications for global governance

The implications of this perspective for global governance are stark and unpalatable. During the 1970s the prospect of biophysical limits to growth was dismissed by the left in the West because the Keynesian social compact was depended on growth. It was dismissed in the Global South because development depended likewise not only on endogenous growth, but access to lucrative markets in the advanced industrial nations. Nothing has changed in the intervening period. Large-scale shifts of manufacturing capital from North to South consequent upon globalization and neo-liberal free trade policies enforced by institutions such as the World Bank and the International Monetary Fund as well as the priorities of the G7, have produced a very visible transformation in countries across Asia and parts of Africa. The development of domestic high-tech innovation hubs, military capacity, and a burgeoning middle class in cities as far removed as Nairobi, Mumbai, and Shanghai have all underlined the undeniable medium-term advantages of industrial growth and development. There is a deeply entrenched and realistic assessment on the part of governing and corporate elites that social cohesion within and geo-political stability among nation-states that are fragile and still in the process of formation depends on the flow of material and financial benefits that come from economic growth. But at the same time, this Faustian pattern of development feeds the narcissistic "we-identity" and flattering self-image of citizenries that is perhaps a necessary concomitant of an effective "imagined community" (Anderson 1991).

How then to break the cycle? Nation-states, trade organizations, corporate lobbies, international institutions, NGOs, consulting companies and the great majority of academic experts are wedded to a pattern of growth that, by definition, cannot deliver sustainable development; and to development goals the rigorous pursuit of which will precipitate or accelerate a trajectory of economic and ecological collapse. But this said, there are very clear first steps – some of which are in fact concordant with current thinking in relation to sustainable development; and others which are gently subversive of this agenda.

First, international agencies and other organizations (including academics) should be more consistent about using alternative indicators other than GNP/ GDP in order to open up a wider civic conversation among general publics, North and South, as to the overarching societal goals. Such simple metrics are very significant in structuring development discourse and investment priorities and all public opinion in all nation states is overly receptive to messages about relative national standing in international league tables (whether football, economic growth, innovation, etc.)

Second, where possible research and investment should be redirected to areas of technical, economic, and regulatory innovation which have the potential radically to reduce the unit "transformity cost" of given products and services, i.e. the embedded energy associated with the distributed web of activities across the entire production system that are brought to bear in the production, delivery, and consumption of any particular item (Odum 2007). This would include, for instance: open source innovation platforms; and micro-fabrication systems facilitating radical localization of economic activity (e.g. 3D printing). It would also include regulatory changes that invert the unit cost of regulation so that the burden is proportional to the size of enterprises, their scope and geographical scale of operations. The impact of such changes would be to unleash the power of the market in local/familial and community contexts whilst raising the cost of intra-regional and international trade and globalized production systems. Kevin Carson's (2010) "Homebrew Industrial Revolution" gives some indication of what such a political economy might look like – libertarian with respect to the state; but communitarian with respect to relations between people in particular places.

Third, any limits perspective must at some point address the problem of global trade. Modernity has been built upon a foundational commitment to the theory of comparative advantage. But in a world of limits and zero sums, this "empty world" perspective begins to break down in the face of the more traditional mercantilist and Malthusian concern with absolute advantages. Tim Morgan (2018) has argued that the protectionism that is beginning to appear in the wake of the Trump victory represents a new, albeit perhaps unwitting, kind of economic realism. Certainly, from the perspective of EEGs, it should be assumed as a matter of priority that the goal of public policy should be to reduce levels of trade, to shorten production chains, and to relocalize production in regional blocs, national economies, and local communities. At the level of discourse, EEGs would side unequivocally against both the corporate agenda of globalization but also the liberal cultural commitment to globalism – and even the idea of global governance. There is a paradox here in that it is reasonably assumed by the great majority of commentators that global governance is a prerequisite for addressing planetary problems such as climate change. The usually unacknowledged assumption is that such a commitment will necessarily be rooted in the intensifying interdependence of institutions, corporations, policy-makers, opinion formers, and cultural elites – communities of thought and practice that can affect the necessary regulatory restraints and co-operation to solve global problems. This makes complete sense within the framework of sustainable development, ecological modernization and SDGs. But it makes no sense in a world of limits. At a most basic level, the order of magnitude increase in complexity associated with this frenetically integrating world – the 350 million products offered through Amazon; the constant escalation of air travel etc. – must entail a corresponding increase in the metabolic load, energy flows, pollution, and resource consumption. And the same pattern of functional interdependence and geographical dispersal of production and consumption activities also creates ever more insuperable barriers to the flows of information in the system, impeding any checks and balances and creating negative feedback loops. A renewed emphasis on the integrity of nation states and national economies combined with sub-national localization is a pre-requisite for addressing what Meadows (2008) argued were key leverage points in the process of system change, namely the flow of information and the control of feedback loops. At the same time, reducing international trade flows and interdependencies whilst re-centring the nation state as well as subnational communities, would also address the third most significant in her list, namely the distribution of power over the rules of the system, by prioritizing the principle of economic and political *subsidiarity* (Médaille 2010).

Finally, participants in development discourse and policy should begin to be much more circumspect about advancing taken-for-granted shibboleths, categories, and goals of Enlightenment social-emancipatory thinking. Instead, the emphasis should routinely focus on wicked dilemmas, paradoxes, and trade-offs – and on the possibility that "better" might be good enough and more sustainable than "perfect". Also, and even more unsettling, there should be a focus on the prospect that completely different, traditional approaches to questions of moral regulation, social cohesion, and culturally embedded understandings of progress may be more appropriate or viable than the shopping list that features in the relevant UNESCO/UNHCR or Oxfam operations manual. A case in point would be a more tempered understanding of the relationship between individual rights and familial/community identities and obligations. This is by no means a recommendation to abandon the concept of human rights, but rather to recognise that effective human restraint, conscience formation, and non-state/market forms of mutual aid and care may require a different or additional ontological framework, and that any clear-cut set of values and categories carries with it a cost. This emphasis on the deep categories at work in the development narrative addresses what Meadows (2008) argued were the two most central levers for system change – the goals of the system and the mindset or paradigm out of which the system, its goals, power structure, rules, and its culture arises.

References

Alexander, C. 2005. *The Nature of Order: An Essay on the Art of Building and the Nature of the Universe.* Books 1 to 4. Berkeley: Center for Environmental Structure.

Alexander, C., H. Neis, A. Anninous, and I. King. 1987. *A New Theory of Urban Design.* Oxford: Oxford University Press.

Anderson, B. R. O. 1991. *Imagined Communities: Reflections on the Origin and Spread of Nationalism.* Revised and extended ed. London: Verso.

Bakker, J. 1990. "The Gandhian Approach to Swadeshi or Appropriate Technology: A Conceptualization in Terms of Basic Needs and Equity". *Journal of Agricultural Ethics* 3 (1): 50–88.

Bauhardt, C. 2014. "Solutions to the Crisis? The Green New Deal, Degrowth, and the Solidarity Economy: Alternatives to the Capitalist Growth Economy from an Ecofeminist Economics Perspective". *Ecological Economics* 102: 60–68.

Bauman, Z. 2000. *Liquid Modernity*. Cambridge: Polity Press.

Beck, U. 1992. *Risk Society: Towards a New Modernity*. London: SAGE.

Carson, K. 2010. *The Homebrew Industrial Revolution: A Low-Overhead Manifesto*. North Charleston, SC: Book Surge.

CHF-BRI (Canadian Hunger Foundation and Brace Research Institute). 1983. *A Handbook on Appropriate Technology*. 3rd edn. Ottawa: Canadian Hunger Foundation.

Daly, H. E., and J. Farley. 2011. *Ecological Economics: Principles and Applications*. 2nd ed. Washington, DC: Island Press.

Dickinson, J. 2009. "The People Paradox: Self-Esteem Striving, Immortality Ideologies, and Human Response to Climate Change". *Ecology and Society* 14 (1): 34.

Donnellan, A. B., K. H. Trzesniewski, and R. W. Robins. 2009. "An Emerging Epidemic of Narcissism or Much Ado about Nothing?". *Journal of Research in Personality* 43 (3): 498–501.

Dreher, R. (2017). *The Benedict Option: A Strategy for Christians in a Post-Christian Nation*. New York: Sentinel.

Durkheim, E. 2006. *On Suicide*. London: Penguin Books.

Elias, N. (1989) 2007. *Involvement and Detachment*. The Collected Works of Norbert Elias. Vol. 8. Dublin: UCD Press.

Elias, N. (1987) 2010. *The Society of Individuals*. The Collected Works of Norbert Elias. Vol. 10. Dublin: UCD Press.

Gellner, E. 1983. *Nations and Nationalism*. Ithaca: Cornell University Press.

Giddens, A. 1991. *Modernity and Self-identity: Self and Society in the Late Modern Age*. Cambridge: Polity Press.

Griggs, D., M. Stafford-Smith, O. Gaffney, J. Rockström, M. C. Öhman, P. Shyamsundar, W. Steffen, *et al.* 2013. "Sustainable Development Goals for People and Planet". *Nature* 495: 305–307.

Hazeltine, B., and C. Bull. 2003. *Field Guide to Appropriate Technology*. Boston, MA: Academic.

Heer, D., and A. Grossbard-Shechtman. 1981. "The Impact of the Female Marriage Squeeze and the Contraceptive Revolution on Sex Roles and the Women's Liberation Movement in the United States, 1960 to 1975". *Journal of Marriage and Family* 43 (1): 49–65.

Illich, I. 1973. *Tools for Conviviality*. Berkeley, CA: Heyday Books.

Jacobs, J. (1978) 2011. *The Death and Life of Great American Cities*. 50th anniversary ed. New York: Modern Library.

Katz, L., and C. Goldin. 2002. "The Power of the Pill: Oral Contraceptives and Women's Career and Marriage Decisions". *Journal of Political Economy* 110 (4): 730–770.

Kish, K. 2018. "Ecological Economics 2.0: Ecological Economic Development Goals". PhD diss., University of Waterloo.

Kish, K., and S. Quilley. 2017. "Wicked Dilemmas of Scale and Complexity in the Politics of Degrowth". *Ecological Economics* 142: 306–317.

Kish, K., J. Hawreliak, and S. Quilley. 2016. "Finding an Alternate Route: Towards Open, Eco-Cyclical, and Distributed Production". *Journal of Peer Production* 9 (9), September.

Accessed 24 November 2018. http://peerproduction.net/issues/issue-9-alternative-internets/peer-reviewed-papers/finding-an-alternate-route-towards-open-eco-cyclical-and-distributed-production/.

Loyal, S. and Quilley, S. 2017. "The Particularity of the Universal: Critical Reflections on Bourdieu's Theory of Symbolic Power and the State". *Theory and Society* 46 (5): 429–462.

Mantena, K. 2012. "On Gandhi's Critique of the State: Sources, Contexts, Conjectures". *Modern Intellectual History* 9 (3): 535–563.

Meadows, D. 2008. *Thinking in Systems: A Primer*. London: Earthscan.

Médaille, J. C. 2010. *Toward a Truly Free Market: A Distributist Perspective on the Role of Government, Taxes, Health Care, Deficits, and More*. Wilmington, NC: Intercollegiate Studies Institute.

Milbank, J., and A. Pabst. 2016. *The Politics of Virtue: Post-liberalism and the Human Future*. London: Rowman & Littlefield International.

Mol, A., and G. Spaargaren. 2000. "Ecological Modernisation Theory in Debate: A Review". *Environmental Politics* 9 (1): 17–49.

Morgan, T. 2018. "#133: An American Hypothesis. Is Donald Trump the First 'Economic Realist'?". *Surplus Energy Economics*, 24 August. Accessed 1 October 2018. https://surplusenergyeconomics.wordpress.com/2018/08/24/133-an-american-hypothesis/.

Neuwirth, R. 2006. *Shadow Cities: A Billion Squatters, a New Urban World*. London: Routledge.

Odum, H. T. 2007. *Environment, Power, and Society for the Twenty-First Century: The Hierarchy of Energy*. New York: Columbia University Press.

Ollman, B. 1977. *Alienation: Marx's Conception of Man in Capitalist Society*. 2nd edn. Cambridge: Cambridge University Press.

O'Neill, D. W., A. L. Fanning, W. F. Lamb, and J. K. Steinberger. 2018. "A Good Life for All within Planetary Boundaries". *Nature Sustainability* 1: 88–95.

Ostrom, E. 1990. *Governing the Commons*. Cambridge: Cambridge University Press.

Pinker, S. 2012. *The Better Angels of our Nature: Why Violence has Declined*. New York: Penguin Books.

Polanyi, K. 1944. *The Great Transformation. The Political and Economic Origins of Our Time*. Boston: Beacon Press.

Quilley, S. 2012. "System Innovation and a New 'Great Transformation': Re-Embedding Economic Life in the Context of 'De-Growth'". *Journal of Social Entrepreneurship* 3 (2): 206–229.

Quilley, S. 2015. "Resilience through Relocalization: Ecocultures of Transition". In *Ecocultures: Blueprints for Sustainable Communities*, edited by S. Böhm, Z. P. Bharucha, and J. Pretty, 199–217. London: Routledge.

Quilley, S. 2017. "Navigating the Anthropocene: Environmental Politics and Complexity in an Era of Limits". In *Handbook on Growth and Sustainability*, edited by P. A. Victor, and B. Dolter, 439–470. Cheltenham: Edward Elgar Publishing.

Quilley. S. (2019) "Individual or Community as a Frame of Reference for Health in Modernity and in the Anthropocene". In *Health in the Anthropocene: Living Well on a Finite Planet*, edited by S. Quilley, and K. Zywert. Toronto: Toronto University Press.

Rifkin, J. 2009. *The Empathic Civilization: The Race to Global Consciousness in a World in Crisis*. New York: J.P. Tarcher/Penguin.

Rockström, J., W. Steffen, K. Noone, A. Persson, F. S. Chapin III, E. F. Lambin, T. M. Lenton, *et al.* 2009. "A Safe Operating Space for Humanity". *Nature* 461 (7263): 472–475.

Schumacher, E. 1989. *Small is Beautiful: Economics as if People Mattered*. San Bernardino, California: Borgo Press.

Scott, J. 1998. *Seeing Like a State: How Certain Schemes to Improve the Human Condition have Failed*. New Haven, CT: Yale University Press.

Turner, J. F. C. 1976. *Housing by People*. London: Marion Boyars.

Twenge, J M. 2017. *IGen: Why Today's Super-Connected Kids are Growing up less Rebellious, More Tolerant, Less Happy – and Completely Unprepared for Adulthood (and What this Means for the Rest of Us)*. New York, NY: Atria Books.

Twenge, J. M., and J. Foster. 2008. "Mapping the Scale of the Narcissism Epidemic: Increases in Narcissism 2002–2007 within Ethnic Groups". *Journal of Research in Personality* 42 (6): 1619–1622.

Twenge, J. M., and Campbell, W. 2010. *The Narcissism Epidemic: Living in the Age of Entitlement*. 1st edn. New York: Free Press.

UNGA (United Nations General Assembly). 2015. *Transforming Our World: The 2030 Agenda for Sustainable Development*. A/RES/70/1. Accessed 17 November 2018. https://undocs.org/A/RES/70/1.

UNWGIP (United Nations Working Group on Indigenous Peoples). 1994. *Draft Convention on the Rights of Indigenous Peoples – Submitted to the Sub-commission on the Prevention of Discrimination and Protection of Minorities*. Accessed 7 December 2018. https://en.wikisource.org/wiki/Draft_United_Nations_Declaration_on_the_Rights_of_Indigenous_Peoples#Article_7.

Ward, C. 1976. "The Do-It-Yourself New Town". *Ekistics* 42 (251): 205–207.

Waugaman, R. 2011. "The Narcissism Epidemic, edited by Jean W. Twenge & W. Keith Campbell". *Psychiatry: Interpersonal and Biological Processes* 74 (2): 166–169.

Weber, M. 1968. *Economy and Society; An Outline of Interpretive Sociology*, edited by Guenther Roth and Claus Wittich. New York: Bedminster Press.

Zhang, E., A. Kleinman, and W. Tu. 2011. *Governance of Life in Chinese Moral Experience: The Quest for an Adequate Life*. Abingdon: Routledge.

Zywert, K., and S. Quilley. 2017. "Health Systems in an Era of Biophysical Limits: The Wicked Dilemmas of Modernity". *Social Theory and Health* 16 (2): 188–207.

Zywert, K., and S. Quilley, eds. 2019. *Health in the Anthropocene: Living Well on a Finite Planet*. Toronto: Toronto University Press.

12 Development as usual

Ethical reflections on the SDGs

Seyed Ali Hosseini

Introduction

Ethics are an important aspect of the Sustainable Development Goals (SDGs), the biggest and most comprehensive global development initiative to combat hunger, poverty, illiteracy, communicable diseases, corruption, and other problems to date. Yet, while ethical concerns are apparently a foundational aspect of the SDGs, they have not attracted enough critical scholarly attention, especially from non-dominant ethical perspectives. Since supposedly "public policies are driven by moral commitments" (Astroulakis 2011, 220) and since ethical debates are at the heart of the legitimacy of any discourse with the power to motivate individuals and institutions to act (Salamat 2016), it is crucial to examine the ethical assumptions of SDGs.

This chapter offers a critical engagement with ethical aspects of the foundations and ideals of the SDGs. It examines the moral challenges that this global initiative seeks to overcome, the ethical approach of the SDGs, and their quite profound limitations based as they are in the global development agenda. The main criticism in this chapter is that the ethics of the SDGs are not different from the ethics governing the previous development thinking that was hegemonic in the last 40 years. Since the efforts to instigate a new international economic order were defeated in the 1970s, the SDGs suffer from the same shortcomings seen before. Such weaknesses limit the SDGs in changing the harmful behavioural patterns of development by addressing only side effects of hegemonic developmental thinking.

The analysis that follows draws on scholarly discussions of development ethics and global development literatures. It starts with some reflections on classical definitions of a good life which emphasize different aspects of human life that are frequently subsumed under neoliberal economic assumptions. The analysis also considers technocratic practices that presume that instrumental actions necessarily produce ethically beneficial outcomes. The final section considers some of the global governance implications of a more ethically encompassing agenda that involves human aspirations based on more than what mainstream notions of development have so far incorporated. The key point made is that, despite the aspirational statements in the SDGs and their justification in *The Future We Want*, the instrumental economic approach to their implementation leaves much to be desired if a larger definition of a good life is taken seriously.

A classical definition of good life

The implicit assumption of development thinking is that a good life is achieved through the qualitative enrichment of all aspects of human life. It promotes efforts for improving living conditions for both individuals and the global community (Astroulakis 2011). But development thinking focuses mostly on the need for economic growth. If a good life is to be understood in larger terms, as more than a matter of the utility maximization that economic growth supposedly delivers, then growth of things other than the economy needs to be explicitly considered. Some of the SDGs are directly related to the good life and well-being of people, including Goal 3 on health, Goal 4 on quality education, Goal 6 on access to clean drinking water and sanitation, Goal 7 on access to sustainable energy, and Goal 11 on creating inclusive human settlements. But given the patterns of global inequality, the widespread excesses of consumption while billions live in poverty despite decades of policies of growth, ethical engagement with the goals and global governance requires a consideration of more than conventional economics.

Some scholars' non-economic-growth approach to development focuses on its ethical aspects by giving importance to human well-being beyond economics. A major effort in this direction is Amartya Sen's and Martha Nussbaum's conception of the capability approach, a departure from hegemonic utilitarianism in economics, as it includes concerns of quality of life, well-being, agency, social justice, freedom, and rights. Instead of an emphasis on goods, consumption, or satisfaction, Sen defines a good life as a productive and creative life according to the needs, values, and interests of the people concerned (St. Clair 2006). Sen considers the dual roles of human beings as agents, beneficiaries, and adjudicators of progress as well as being the means of such progress, highlighting the grounds for confusion between ends and means in policymaking. He proposes a notion of development built on human capabilities. This conceptual framework assisted the United Nations Development Programme (UNDP) to compile the first Human Development Report in 1990 (Clark 2002).

This approach is supported by a survey conducted on the aspects of a good life, which indicated that many people consider access to jobs, housing, education, adequate and regular income, having a good family, living religiously, health, food, happiness, and love as aspects of a good life. Such findings place emphasis on the psychological and mental aspects of a good life that most international policymakers and many researchers neglected until recently. These elements include happiness, pleasure, excitement, joy, relaxation from stress, confidence, self-respect, and experiencing pride (Clark 2002). The SDGs' approach to a good life considers provision of needed goods and facilities but neglects psychological and mental needs for a dignified life (see also Freston's [Chapter 10] discussion of a dignified life as seen by different faiths).

Considering the problems produced by the economic, modernist, and indus-trialized notion of development, including depletion of natural resources, huge inequalities, and continued health problems and poverty, it is now the task of development ethics to determine the components of well-being and examine how these might be promoted (St. Clair 2014; Astroulakis 2011). The SDGs' approach, although broader than recent global development agendas, still focuses on the economic notion of development, disregarding many factors of a good life and well-being.

Ethicists suggest putting moral issues at the heart of the economy. As indi-cated by Rolston (2015, 351),

> economists, like the ecologists, may help tell us what our options are, what will work and what will not. But there is nothing in economics per se that gives economists any authority or skills at making these further social deci-sions about the deeper goals of sustainability.... We must have an ethics that asks about how to live justly.

Some go even further by questioning the domination of international develop-ment by economists who brandish economics as a value-free science with solu-tions for global poverty – an example of that was seen in the revival of the free market as prescribed by the Washington Consensus. The claim of neutrality of economics has been criticized as theory choice is not "based objectively on non-controversial criteria" (Wilber 2010, 157) but on value-laden criteria which serve a goal, like assumptions that human beings always act out of self-interest or that individuals have a rational freedom to pursue happiness as they define it.

A response to human sufferings and degrading earth systems

There are many challenges to people's lives and earth systems in all corners of the globe, some reaching high levels of human suffering and irreversible damages to life systems. These miseries are present even though human wealth and productivity have reached an astonishing level of $127 trillion of output per year, and the average output per person of the world economy is $16,770 (Annett *et al.* 2017, 2). Human knowledge and innovation continue to grow and open new horizons for humanity. Considering such capacities, the current situation of our world is in contradiction with any system of beliefs and any moral values deemed essential for a social life, which include respectful treat-ment of others, human dignity, equality, justice, friendship, and honour (Astroulakis 2011; Motilal 2015). Most of these problems are attributable to a lack of will among the main global actors rather than to a lack of means, and this situation constitutes a major moral problem for humanity.

The unpleasant situation of the world and its inhabitants is at the centre of the SDGs' justification, which is similar to the ethical justification of the previous international development initiatives motivated by a picture of a world of poverty and inequality (Litonjua 2013). The evils of extreme poverty and hunger, the top two issues of all SDG documents, are moral challenges that the conscience of our wealthy world cannot ignore. The outcome document of the UN Conference on Sustainable Development in Rio de Janeiro indicates that "Poverty eradication is the greatest global challenge facing the world today and an indispensable requirement for sustainable development. In this regard, we are committed to freeing humanity from poverty and hunger as a matter of urgency" (UNGA 2012, 2). Perhaps "the most pressing issue of global justice", extreme poverty "is morally unacceptable as it is debilitating and dehumanizing, to the point that individuals are unable to pursue their own ends, flourish or enjoy basic rights" (Moellendorf and Widdows 2015, 152). Poverty is associated, directly or indirectly, with hunger, illiteracy, disease, inequality, unemployment, violence, and corruption. Desperation caused by poverty often brings people to sacrifice long-term benefits for short-term survival needs or even to engage in violence. Extreme poverty produces extreme problems, which in turn perpetuate the vicious cycles.

In addition to human suffering, there are man-made damages, some irreversible, to earth systems, including climate change, loss of biological diversity, depletion of the ozone layer, pollution, and exhaustion of resources, including water. For some ethicists, climate change vies with the immorality of poverty to form the two most important moral challenges of our generation. The immorality of this man-made phenomenon lies in its potential to render large parts of the earth uninhabitable, to endanger ecosystems and human life (St. Clair 2006). Additionally, preservation of biodiversity is a moral and aesthetic responsibility: biodiversity reflects the cultural diversity of human beings, since many indigenous communities live in areas of mega-biodiversity and their cultures have been fashioned by biodiverse nature (UNEP 1999).

The moral responsibility of global actors to address the suffering of people and to keep the earth habitable for future generations was the main motivator for the creation of the SDGs. Depicting a vision of free, fair, and full world is promising, as the UN resolution founding the SDGs indicates:

> In these Goals and targets, we are setting out a supremely ambitious and transformational vision. We envisage a world free of poverty, hunger, disease and want, where all life can thrive. We envisage a world free of fear and violence. A world with universal literacy ... food is sufficient, safe, affordable and nutritious. A world where human habitats are safe, resilient and sustainable....
>
> (UNGA 2015, 3–4)

These concerns motivated global actors to design the global development initiatives of the SDGs upon the conclusion of the Millennium Development Goals (MDGs).

Continuation of the previous development programs

The SDGs built on the MDGs and other international development agendas and followed their assumptions and foundations with frequent references to the previous UN documents, principles, and conference outcomes. As the founding document of the SDGs indicates, the 17 SDGs and their 169 targets "seek to build on the Millennium Development Goals and complete what they did not achieve" (UNGA 2015, 1), even though they are comprehensive and follow broad consultancy with a range of actors. The SDGs are "guided by the purposes and principles of the Charter of the United Nations, with full respect for international law and its principles" (UNGA 2012, 1), and they "envisage a world of universal respect for human rights and human dignity, the rule of law, justice, equality, and non-discrimination" (UNGA 2015, 4).

Basing the SDGs on human rights is promising since, according to many scholars, different cultures and worldviews converge around the notion of human dignity and equality for all (Mele and Sánchez-Runde 2013). Human rights provide development practitioners with moral values as well as enforcement power to develop human capabilities and to protect specific groups including workers, women, and children against potential abuses in business and development (Mele and Sánchez-Runde 2013; Forst 2015). As "universal valid moral norms" (Enderle 2014, 166), human rights include a range of entitlements for individuals, including the right to development. Many of the SDGs are posed as rights (e.g. the right to development, the right to adequate standard of living, and the right to food and rights to education) and they have a moral basis in human rights.

Aside from this advantage, linking the enjoyment of the SDGs to the human rights agenda may pose some limitations. Regarding the moral aspect, human rights are seen as "minimal moral norms and do not encompass all moral norms and values" (Enderle 2014, 166). Also, implementation of human rights would entail partial realization of global ethics and global justice, as Dower (2014, 12) depicted the relationship between human rights, global ethics, and global justice as a triangle with some overlapping areas. Human rights are not necessarily effective when it comes to social, economic, and cultural claims against states. Additionally, they can hardly be effective against other states, companies, and other actors.

Absence of political structures of peace

The SDGs recognize the mutual effect of peace and development as "There can be no sustainable development without peace and no peace without sustainable development" (UNGA 2015, 2), and that the SDGs agenda "seeks to strengthen universal peace in larger freedom.... We are resolved to free the human race from the tyranny of poverty and want and to heal and secure our planet" (UNGA 2015, 1). The SDGs' combat against poverty, hunger, illiteracy, discrimination, and corruption is supposed to serve peace in the long run.

This approach reflects development scholars' findings on the mutual causal effects of war and violence, linked to problems of hunger, ignorance, exclusion, and poor health conditions.

Against the plausibility of a development approach to peace, it is not established that current development thinking brings peace to our world. While war and violence are not detached from politics, the SDGs' triangle of economic, social, and environmental development does not provide any space for politics, neglecting the political enablers of problems, such as war, state fragility, and migration. Most of the lingering problems targeted by the SDGs are present in the societies that have been entangled in war or violence, and the SDGs, as a depoliticized agenda, can do little to emancipate those societies. If implemented well, the SDGs can sustain peace mainly in peaceful developing countries. While SDG 16 aims at promoting peace and justice in all communities, it failed to address war, which is the main cause of extreme forms of poverty, hunger, unemployment, and other problems in different parts of the world. While improvement of elements of good government, including accountability, participation, inclusiveness, rule of law, and transparency, contribute to the prevention of war and mass violence, there are also political aspects, such as foreign intervention, that play a part in the outbreak and continuation of war and violence in fragile and failed states. This issue merits addressing by global governance, including through improved attention to the ethical aspects of governance (see Dalby, Chapter 8).

Conflicts and wars are preventable, and not inevitable, if global politics is corrected. Just to give a clear example, as long as the UN Security Council's permanent members veto resolutions addressing conflicts, war, or international crime in different parts of the world, such phenomena continue to destroy structures that are necessary for income, education, health, and safe societies. The humanitarian assistance and partnership envisaged in the SDGs are just reliefs from structural malfunction. Therefore, global development governance cannot ignore the "weak, imprecise, or ineffective" norms governing global politics, including the UN's undemocratic charter (Aydinalp 2016, 209–210), or other negative political attitudes in regional levels. But these political problems are compounded by the economic assumptions that underlie the development project as it has been structured since the 1980s.

Centrality of the neoliberal economic system

A crucial part of the problem with the neoliberal approach is its de-politicization of economics and development by reducing them to technical issues (Best 2006). The de-politicization of development is not limited to the political economy of development agendas; it also leads to political unawareness among practitioners (Malavisi 2014). Considering this fact, the SDGs could hardly make a difference, since the main global actors of politics and the economy are the same as before. Given the different and even competing values, goals, and ways of operation among businesses, governments, donors, financial institutions, and civil society actors, the SDGs agenda has been a compromise and hence it is unlikely to deal with the economic roots of current problems in unjust structures (UNGA 2012; Scheyvens, Banks, and Hughes 2016). Thomas Pogge refers to the structural injustice in the global institutional order, including the International Monetary Fund (IMF), World Bank (WB), and World Trade Organization (WTO), as a main cause of global poverty (Malavisi 2016, 185). At the same time, many NGOs that claim to have a distinct and egalitarian perspective on development receive funding from those organizations and are part of the same failing global system (Malavisi 2016). This view takes us to the belief that the de-politicization of development reduces it to tackling the side-effects of a failing global system, rather than delivering the transformation that its rhetoric promises.

Economic growth is one of the three pillars of the SDGs initiative. SDG 8 aims at the promotion of inclusive, sustained economic growth and of full employment, while SDG 9 promotes building an inclusive industrialization and fostering innovation. SDG 17 (Targets 10, 11 and 12) promotes "a universal, rules-based, open, non-discriminatory and equitable multilateral trading system under the World Trade Organization"; it aims to "Realize timely implementation of duty-free and quota-free market access on a lasting basis for all least developed countries, consistent with World Trade Organization decisions" (UNGA 2015, 27). This approach has been dominant in global development initiatives in the post-Cold-War era and it is considered inherent to development.

Such a reliance on economic growth dovetails with the widespread conception that people are motivated by self-interest and that, when there is no personal benefit at stake, people will not stand for change (Salamat 2016). Accordingly, international development has been perceived mainly in terms of straightforward economic production, and all policymakers of development have considered economic growth, technological expansion, and engineering nature to be at the heart of any development agenda (Astroulakis 2011). To attain such a level of economic growth and capital accumulation, modern development thinking promotes technological change, structural transformation of the economy, as well as modernization of institutions, industrialization, consumerism, and the use of natural resources as development (Clark 2002; St. Clair 2014).

For the ethicists, consumption is "the mother of all environmental issues" (Isenhour 2015, 133), especially consumption of non-renewable resources. Overconsumption raises critical ethical questions because it is considered unjustifiable due to the scarcity of some resources, its increasing inequality, and the fact that there is a moral responsibility to avoid terrible choices (Moellendorf and Widdows 2015). Connecting this moral issue to development, inequality is at the heart of the current hegemonic lifestyle as, for example, a study by Oregon State University in 2009 concluded that on average a child in the US has "a long-term effect on raising carbon emissions that is 160 times higher than a child in Bangladesh" (Moellendorf and Widdows 2015, 369). Yet, Western lifestyle has been reflected in sustainable development, which focuses on economic growth, technology, and the implicit assumption that this mode of living is a universal aspiration.

The SDGs rely on the market economy and economic growth-based development, only to be improved by more complementary environmental preservation and neoliberal social justice. A neoliberal economy is clearly at the heart of the economic growth that the SDGs seek to achieve:

> We reaffirm that international trade is an engine for development and sustained economic growth, and also reaffirm the critical role that a universal, rules-based, open, non-discriminatory and equitable multilateral trading system, as well as meaningful trade liberalization, can play in stimulating economic growth and development worldwide, thereby benefiting all countries at all stages of development as they advance towards sustainable development.
>
> (UNGA 2012, 71)

For ethicists, the global economic system is driven by a "constant ambition to maximize profit, accumulate capital, decrease costs, create low-priced labour, generate new areas of demand, foster unregulated market competition, and establish monopolies" (Aydinalp 2016, 210). Ethicists also post that, fuelled by the quest for more profit, the market economy does little for the common good (Annett *et al.* 2017). Competition to feed self-interest is the rule of the market, and market competition determines who is a winner and who is a loser. Evidence shows that an "economic-based self-interest" approach seems to fail to motivate stakeholders to incorporate the social and environmental dimensions in their national and organizational development planning in an efficient and timely manner (Salamat 2016, 4).

Related to the SDGs implementation is the less-regulated and self-regulated market where businesses have increasing influence on the well-being of people. Business leaders hold that economic growth will lead to shared prosperity and that governments should work to create a "business-friendly trade system, pricing incentives, transparent procurement, and to encourage and support responsible business" for private enterprises which deliver on sustainable goals (Scheyvens, Banks, and Hughes 2016, 374). Private-sector presence alongside political leaders and civil society actors in designing the SDGs reflected a conscious shift toward giving this sector more space in designing development policy and planning (Scheyvens, Banks, and Hughes 2016). In turn, the private sector has developed self-committing principles, such as the Caux Roundtable for Moral Capitalism (2010) and International Organization for Standardization (ISO) on "Guidance on Social Responsibility" (also known as ISO 26000), to replace abandoned state regulation on improving the economic and social conditions of human life and contribution to sustainable development. ISO Guidance on Social Responsibility addresses seven core subjects of social responsibility, including accountability, transparency, ethical behaviour, respect for stakeholder interests, respect for the rule of law, respect for international norms of behaviour, and respect for human rights. Aside from problems that may undermine ISO Guidance on Social Responsibility implementation and effectiveness, including excessive broadness, and the fact that it is costly and time-consuming for small business (Hemphill 2013), the ethical problem with this and other private sector initiatives is their self-interest motivation. The motivation behind observation of the guidance and preservation of the social responsibility is clearly materialistic: "An organization's level of social responsibility influences everything from its reputation to its ability to attract high-calibre employees and its relationship with suppliers, clients and the communities in which it operates" (International Organization for Standardization 2016, 4).

Minimalist environment protection

Unsustainable development, with its polluting industries and relentless exploitation of natural resources, has caused environmental damages, some irreversible. According to scientific assessment, a world four or more degrees warmer than the pre-industrial era endangers food systems and would cause flooding, severe weather events, and irreversible biodiversity and ecosystem losses, which in turn produce more displacements, diseases, and insecurity. Such climate change would endanger many human development achievements (St. Clair 2014, 284). Promoting economic development compatible with natural systems is no longer a choice; now it is a necessity (see Dalby, Chapter 8). This necessity is well reflected in the SDGs, including in the UN General Assembly Resolution of *The Future We Want*, which states that "in order to achieve a just balance among the economic, social and environmental needs of present and future generations, it is necessary to promote harmony with nature" (UNGA 2012, 10). *Our Common Future* stresses that "Human laws must be reformulated to keep human activities in harmony with the unchanging and universal laws of nature" (World Commission on Environment and Development 1987, 271). Goals 12, 13, 14, and 15 on sustainable consumption and reproduction, combating climate change, sustainable use of marine sources, and sustainable use of terrestrial ecosystems are dedicated to this cause.

In the language of the United Nations Environmental Programme (UNEP 1999), there should be a balance between economy and ecology, which means stopping anthropogenic harm on land, sea, and air. But this change cannot be achieved by the current market-induced competition and consumption as "when everybody seeks their own good, there is escalating consumption" (Rolston 2015, 351). To halt climate change, we need not only a limit or change in current trends of resource extraction and pollution, but a substantive paradigm shift in development by introducing new concepts of growth and wealth (St. Clair 2014, 287). For such a paradigm shift to happen, borrowing from other approaches of development, including indigenous, biocentrism, and eco-centrism, should be considered.

Harmony with nature and paying special attention to the environment and biodiversity are elements of indigenous ways of life. In the view of many indigenous communities, with 370 million people in over 70 countries, nature is not an economic commodity, and human life is not separate from nature. Rather, nature is understood as part of the cultural heritage of human life (UNPFII 2006; UNEP 1999). Also, the monetary immeasurability of life and non-compensable character of biodiversity have been proclaimed by many indigenous voices (UNEP 1999, 122).

Regarding the other approaches, some scholars suggest that humans should represent non-humans, which form a significant portion of life, in decision-making processes to make sure that their needs are not subordinated to human priorities (Fredericks 2015, 82). This issue is more important when the question of sustainability arises. While current prevailing anthropocentrism emphasizes human values and recognizes animal and nature protection to the extent that they benefit human beings economically, medically, aesthetically or culturally, biocentrism and ecocentrism emphasize the intrinsic value of all living beings (Keithsch 2018; Garner 2015). For those ethicists, "sustainability cannot be ethics-free. It cannot be based on an anthropocentric 'human supremacy' approach, where humanity always seeks to be the 'master'" (Washington 2015, 371, emphasis in the original). Humans are geologically superior creatures on the Earth and they have severely affected earth systems, in some respects to dangerous and irreversible levels through knowledge and technologies accumulated over centuries (St. Clair 2014). This superiority, combined with dangerous activities, provokes a sense of responsibility toward preserving the Earth and its inhabitants.

The SDGs approach recognizes the importance of a change toward nature by indicating that "Sustainable development requires changes in values and attitudes towards environment and development – indeed, towards society and work at home, on farms, and in factories ..." (World Commission on Environment and Development 1987, 95). While such an approach may provide some space for other perspectives, it is still human-centred as it indicates that:

> We reaffirm the intrinsic value of biological diversity, as well as the ecological, genetic, social, economic, scientific, educational, cultural, recreational and aesthetic values of biological diversity and its critical role in maintaining ecosystems that provide essential services, which are critical foundations for sustainable development and human well-being.
>
> (UNGA 2012, 52)

The question of realization of social justice

The issue of justice forms part of any ethical theory (Cortina 2014): social justice and equal distribution of wealth have long been at least a moral demand. There are varying degrees of conceptual cross-over between global ethics, human rights and global justice, although the primary normative focus of each remains different (Dower 2014). Entirely related, equality is at the heart of the justice question and inequality of all kinds is a concern for scholars from different fields. Inequality between countries, classes, races, and sexes is an unpleasant reality with deep roots in social, political, and economic structures that produced and continue to reproduce inequality (Crocker 2014).

Social justice has been among the victims of the modern development approach. There has been a growing concern about well-being of socially disadvantaged groups, including workers, women, children, and migrants, especially following a series of economic crises in the last 40 years. The SDGs recognized the importance of social justice as "promoting universal access to social services", and they "strongly encourage initiatives aimed at enhancing social protection for all people" (UNGA 2012, 29). As a pillar of sustainability, social justice has been represented in different goals, including Goal 5 on gender equality and women empowerment, Goal 10 on reduction of inequality, and Goal 16 on promotion of inclusive societies and access to justice. Target 5.5 speaks of full participation and equal opportunities for women in leadership at all levels of political, economic and public life. Goal 16 affirms the need to "promote peaceful and inclusive societies for sustainable development, provide access to justice for all and build effective, accountable and inclusive institutions at all levels" (UNGA 2015, 14), while Target 16.7 speaks of ensuring "responsive, inclusive, participatory and representative decision-making at all levels" (UNGA 2015, 25).

The SDGs' emphasis on inclusivity can serve as their ethical foundation, as:

> we pledge that no one will be left behind. Recognizing that the dignity of the human person is fundamental, we wish to see the goals and targets met for all nations and peoples and for all segments of society. And we will endeavour to reach the furthest behind first.
>
> (UNGA 2015, 3)

It also promotes inter-generational justice, which includes consideration of the living conditions of future generations, to meet the criteria mentioned in the 1987 World Commission on Environment and Development Report of not "compromising the ability of future generations to meet their own needs" (World Commission on Environment and Development 1987, 16). Here, the centre of the issue is implementation of these goals.

Crocker (2014) believes that to answer the current shortcoming of global development, the issues of inequality of power, agency, and empowerment, among others, should be considered. When the decisions and actions of individuals and groups to make a difference in the world are not compromised by compulsion, they are considered independent agents and empowered. For individuals to be well and do well, they should have a real opportunity to exercise their agency, which means they should have security, food, and other minimum conditions to act in a way that increases their well-being and that enables them to combat obstacles to a better life (Crocker 2014). In my view, the point is the "real opportunity". Many of the promises and pledges of the SDGs, or even those guaranteed by international human rights law, provide only potential opportunities to individuals. Many marginalized people and communities do not have good enough conditions to exercise their agency and employ those opportunities to bring about a positive change in their lives.

Furthermore, there is also a debate on the ethical dimension of the social justice version followed by the SDGs. The redistribution of wealth in the current global economy is under the influence of neoliberalism, which gives priority to the market. The market approach to social justice is a micro-economic remedy for macro-economic failures. Critics object that since neoliberal mechanisms have led to social inequalities and heightened the power imbalance between countries, the same mechanisms should not be relied on for solving inequalities. Additionally, the global goal setting does not focus on the structural causes of poverty, including "illicit financial flows, debt, unfair trade rules and corporate power" (Scheyvens, Banks, and Hughes 2016, 377). If they do not address these key sources of global problems, it is not clear how neoliberal market solutions will affect beneficial change in the lives of those suffering from inequality, marginalization, and an absence of solidarity.

Absence of universal solidarity

The last Goal is dedicated to global partnership between different stakeholders, and it is essential for achieving the ambitious and broad-ranging SDGs. Since ethics are not only concerned with the ends of human action but also with the value aspects of the instruments used for actions (Marangos and Astroulakis 2009), the implementation of global partnership for the SDGs is also a subject for ethical scrutiny.

Goal 17 entails setting up a global partnership that is necessary for the implementation of the previous goals. Beyond being an invitation for joint work, Goal 17's targets set commitments and procedures for partnership:

> Developed countries to implement fully their official development assistance commitments, including the commitment by many developed countries to achieve the target of 0.7 per cent of gross national income for official development assistance (ODA/GNI) to developing countries and 0.15 to 0.20 per cent of ODA/GNI to least developed countries; ODA providers are encouraged to consider setting a target to provide at least 0.20 per cent of ODA/GNI to least developed countries.
>
> (UNGA 2015, 26)

Still, there is a doubt about the sufficiency of such commitments and arrangements. There have been some measures, from aid to technology transfer, to assist developing countries to achieve some of the SDGs, yet many of these measures struggle with their own problems (Litonjua 2013).

Preferential option for the poor is not sufficient, as study after study shows that the quest for economic growth failed to eradicate poverty and inequality, if it did not exacerbate it. From an ethical perspective, there is a need for solidarity far beyond partnership. Solidarity can serve as an ethical strategy (Astroulakis 2011) by producing an environment of common support among all participants, especially since it is supposed that the implementation of the SDGs is everybody's task. In support of universal solidarity, it is said that, at a philosophical level, there is an agreement that people, in spite of all of their differences, have a common "human-ness". Additionally, the habitat of humans, the Earth, is governed by physical laws which influence humans and other living organisms equally. Common destiny is used as another reason for such solidarity (Astroulakis 2011, 227). Solidarity with the poor is only achieved if the well-to-do are able to imagine themselves in the situation of the poor, breaking with ideas of division and superiority. Through solidarity, we can understand the sameness of humanity, encompassing reach and poor. However, "without the virtue of solidarity, the poor are 'they,' the objects of our pity and generosity." (Litonjua 2013, 93). With the latter attitude, "it is almost impossible to avoid being paternalistic and manipulative. In solidarity the poor are 'we,' sensitive to the feelings of others, and devoted to their common welfare" (93). The instrumentalities of neoliberalism frequently fail to engage an ethics of inclusion, one of a common humanity beyond the narrow assumptions of consumer aspiration which sees life's goals and social life as participation in markets.

Conclusion: ethics, development and governance

The history of development indicates that poverty, inequality, environmental degradation, and other problems are caused by economic growth hegemony combined with its moral legitimization and justification, as well as by a lack of moral awareness for consequences of that development (St. Clair 2006). Current global development thinking follows the same economic growth-based version with its dominant individualist neoliberalism, overconsumption, and depoliticization of development. Regarding social justice, this development approach encourages aid instead of solidarity and deals with degrading earth systems and climate change from an anthropocentric perspective.

These specificities of the SDGs merely confine this global development agenda to a basic human moral obligation just to respond to the side-effects of problems created by global systems, rather than a paradigm shift from hegemonic global developmental thinking. It is possible to claim that the SDGs provide a universal minimal morality, just as other minimalist values serve the survival of the planet and people (Mele and Sánchez-Runde 2013). The problem with the universal minimalist morality is its limits and the unsustainability of the development based upon it. This responsive approach conflicts with the fact that reflections on tackling global poverty and inequality require reflections on global systems that produce them (Aydinalp 2016). To transform the world to a safe, free, and environmentally sustainable one, there is a need for ethical revision of global development.

Repeated failures of development schemes have facilitated the emergence of new development ethics as an area in development studies to reflect on the aims, values, and means of the current development approach with the hope of enabling development to address the causes of war and violence (Cortina 2014). While implementation of the SDGs remains vital, significant improvement of global development requires addressing weaknesses in current developmental thinking. Promotion of a more inclusive ethics of development and the economy, as well as connecting development to global politics, through addressing political aspects of war, violence, and migration, and especially the shortcomings of global economic system, can significantly assist the SDGs in reaching their intended aims.

To enhance global development governance there is a need for broader presence of ethics at the definition of development as well as development knowledge. Enhancing global partnership to revitalize global development, reflected in Goal 17, is good but not a paradigm shift. Measures suggested in Goal 17 are necessary in the current global structure, but they are not new. Such measures include the financing of SDGs by developed countries, technology transfer to developing countries, capacity building in developing countries, and favourable trade conditions for developing and less-developed countries. These are positive aspects of usual developmental thinking that, nevertheless, continue to address side-effects of broader failures of economic growth and depoliticized development thinking. To better address the global problems of hunger, poverty, inequality, and injustice, a new approach is needed to define the poor and disadvantaged peoples as part of a global "we". Introduction of broader and new ethics to the global development agenda would entail a broader shift in the global governance of development. Improving global solidarity to encompass every individual and society would work far better than just financial commitment and technocratic partnership.

Global injustices continue to exist because global governance is weak and biased toward the mighty and wealthy. So far, efforts to bring fundamental changes to global governance have not been successful; be it political efforts like UN reform, or economic ones like the attempt to establishment a new international economic order. There is a clear need to think about ethical aspects of global governance in general and global development governance in particular. Such reflections and debates have the potential to introduce academic theories as well as policy mechanisms to significantly improve global governance's mechanisms for addressing global issues. This chapter contributed to some ethical reflections, but there are many issues that can be discussed, including attention to political aspects of global development, debates on the centrality of the economy in development, and non-hegemonic approaches to development, production, and consumption.

References

Annett, A., J. Sachs, M. S. Sorondo, and W. Vendley. 2017. "A Multi-Religious Consensus on the Ethics of Sustainable Development: Reflections of the Ethics in Action initiative". *Economics e-Journal* 2017 (56).

Astroulakis, N. 2011. "The Development Ethics Approach to International Development". *International Journal of Development Issues* 10 (3): 214–232.

Aydinalp, H. 2016. "Reflecting on Global Obstacles for Fighting Global Poverty: Real Politics vs. Real Justice". In *Globality, Equal Development and Ethics of Duty*, edited by M. Masaeli, 206–219. Newcastle upon Tyne: Cambridge Scholars Publisher.

Best, J. 2006. "Co-opting Cosmopolitanism? The International Monetary Fund's New Global Ethics". *Global Society* 20 (3): 307–327.

Caux Round Table for Moral Capitalism. 2010. "Principles for Responsible Business". Accessed 17 November 2018. www.cauxroundtable.org/index.cfm?menuid=8.

Clark, D. A. 2002. "Development Ethics: A Research Agenda". *International Journal of Social Economics* 29 (11): 830–848.

Cortina, A. 2014. "Four Tasks of Forward-Looking Global Ethics". *Journal of Global Ethics* 10 (1): 30–37.

Crocker, D. A. 2014. "Development and Global Ethics: Five Foci for the Future". *Journal of Global Ethics* 10 (3): 245–253.

Dower, N. 2014. "Global Ethics: Dimensions and Prospects". *Journal of Global Ethics* 10 (1): 8–15.

Enderle, G. 2014. "Some Ethical Explications of the UN Framework for Business and Human Rights". In *Sustainable Development: The UN Millennium Development Goals, the UN Global Compact, and the Common Good*, edited by O. F. Williams, 163–183. Indiana: University of Notre Dame Press.

Forst, R. 2015. "Human Rights". In *The Routledge Handbook of Global Ethics*, edited by D. Moellendorf and H. Widdows, 72–81. New York: Routledge.

Fredericks, S. E. 2015. "Ethics in Sustainability Indexes". In *Sustainability: Key Issues*, edited by H. Kopnina and E. Shoreman-Ouimet, 73–87. New York: Routledge.

Garner, R. 2015. "Environmental Politics, Animal Rights and Ecological Justice". In *Sustainability: Key Issues*, edited by H. Kopnina and E. Shoreman-Ouimet, 331–346. New York: Routledge.

Hemphill, T. 2013. "The ISO 26000 Guidance on Social Responsibility International Standard: What Are the Business Governance Implications?". *Corporate Governance* 13 (3): 305–317.

International Organization for Standardization. 2016. *ISO 26000 and the SDGs*. Geneva, Switzerland. Accessed 17 November 2018. www.iso.org/files/live/sites/isoorg/files/archive/pdf/en/iso_26000_and_sdgs.pdf.

Isenhour, C. 2015. "Sustainable Consumption and its Discontents". In *Sustainability: Key Issues*, edited by H. Kopnina and E. Shoreman-Ouimet, 133–155. New York: Routledge.

Keithsch, M. 2018. "Structuring Ethical Interpretations of the Sustainable Development Goals: Concepts, Implications and Progress". *Sustainability* 10 (829): 1–9.

Litonjua, M. D. 2013. "International Development Economics and the Ethics of the Preferential Option for the Poor". *Journal of Third World Studies* 30 (1): 87–119.

Malavisi, A. 2014. "The Need for an Effective Development Ethics". *Journal of Global Ethics* 10 (3): 297–303.

Malavisi, A. 2016. "Development Ethics: Oxymoron or Real Possibility". In *Globality, Equal Development and Ethics of Duty*, edited by M. Masaeli, 183–205. Newcastle upon Tyne: Cambridge Scholars Publisher.

Marangos, J., and N. Astroulakis. 2009. "The Institutional Foundation of Development Ethics". *Journal of Economic Issues* 43 (2): 381–389.

Mele, D., and C. Sánchez-Runde. 2013. "Cultural Diversity and Universal Ethics in a Global World". *Journal of Business Ethics* 116: 681–687.

Moellendorf, D., and H. Widdows, eds. 2015. *The Routledge Handbook of Global Ethics*. New York: Routledge.

Motilal, S. 2015. "Sustainable Development Goals and Human Moral Obligations: The Ends and Means Relation". *Journal of Global Ethics* 11 (1): 24–31.

Rolston, H. III. 2015. "Environmental Ethics for Tomorrow". In *Sustainability: Key Issues*, edited by H. Kopnina and E. Shoreman-Ouimet, 347–358. New York: Routledge.

Salamat, M. R. 2016. "Ethics of Sustainable Development: The Moral Imperative for the Effective Implementation of the 2030 Agenda for Sustainable Development". *Natural Resource Forum* 40: 3–5.

Scheyvens, R., G. Banks, and E. Hughes. 2016. "The Private Sector and the SDGs: The Need to Move Beyond 'Business as Usual'". *Sustainable Development* 24: 371–382.

St. Clair, A. L. 2006. "Global Poverty: Development Ethics Meet Global Justice". *Globalizations* 3 (2): 139–157.

St. Clair, A. L. 2014. "The Four Tasks of Development Ethics at Times of a Changing Climate". *Journal of Global Ethics* 10 (3): 283–291.

UNEP (United Nations Environment Programme). 1999. *Cultural and Spiritual Values of Biodiversity*, edited by D. A. Posey. Intermediate Technology Publications. London. Accessed 17 November 2018. http://wedocs.unep.org/bitstream/handle/20.500.11822/9190/Cultural_Spiritual_thebible.pdf?isAllowed=y&sequence=1.

United Nations General Assembly. 2012. *The Future We Want*. A/RES/66/288. Accessed 17 November 2018. https://undocs.org/A/RES/66/288.

United Nations General Assembly. 2015. *Transforming Our World: The 2030 Agenda for Sustainable Development*. A/RES/70/1. Accessed 17 November 2018. https://undocs.org/A/RES/70/1.

United Nations Permanent Forum on Indigenous Issues (UNPFII). 2006. Indigenous People, Indigenous Voice. Factsheet No. 1. Accessed 17 November 2018. www.un.org/esa/socdev/unpfii/documents/5session_factsheet1.pdf.

Washington, H. 2015. "Is 'Sustainability' the Same As 'Sustainable Development'?". In *Sustainability: Key Issues*, edited by H. Kopnina and E. Shoreman-Ouimet, 359–377. New York: Routledge.

Wilber, C. K. 2010. "Economics and Ethics". In *New Directions in Development Ethics: Essays in Honor of Denis Goulet*, edited by C. K. Wilber and A. K. Dutt, 157–176. Indiana: University of Notre Dame Press.

World Commission on Environment and Development. 1987. *Report of the World Commission on Environment and Development: Our Common Future*. Accessed 17 November 2018. www.un-documents.net/our-common-future.pdf.

13 Financing the Sustainable Development Goals

Beyond official development assistance

Susan Horton

Introduction

Adoption of the Sustainable Development Goals (SDGs) set out an audacious global agenda. From the outset, it was recognized that achieving these goals would require substantial financial resources. Consequently, their adoption was preceded by the Third Conference on Financing for Development in Addis Ababa. This was a contrast to the Millennium Development Goads (MDGs) since the First Conference on Financing for Development in Monterrey only occurred 18 months after the goals' adoption.

At the Addis Ababa meeting, UN Secretary General Ban Ki Moon spoke of "a new era of sustainable development, a paradigm shift towards a truly transformative agenda" (UN Secretary General 2015a) and issued the rallying cry "leave no one behind" (UN Secretary General 2015b). The title of the World Bank side event "From Billions to Trillions" made it clear that official development assistance (ODA) alone could not possibly fund this ambitious agenda.

Using health as an example for the SDG social goals, this chapter examines how the resources to finance the health SDG might be met. It begins with a brief resume of how assistance for international development has evolved, with a focus on the transition from the MDGs to the SDGs. We then review four modalities which are being used to generate innovative financing for the global health agenda. The question of how conventional bank finance might support the SDGs is taken up by Weber (Chapter 14). The concluding section considers some of the global governance issues concerned. We use US dollars throughout, unless specified otherwise.

The theoretical approach taken in this chapter is a fairly pragmatic one, consistent with that taken by Jeffrey Sachs, the UN Secretary-General's Special Adviser on the MDGs (United States). Not all economists support this approach. For example, Easterly (2015) argues that many of the SDGs are variously "unactionable, unquantifiable and ... unattainable" and Angus Deaton expresses concerns as to how they can be measured, particularly the poverty goal (Anders 2016). Non-economists also criticize the UN approach. One of the more extreme viewpoints is Hickel's (2016, 749), which argues that "the UN's claims about poverty and hunger are misleading, and even intentionally inaccurate" and that "our present model of development is not working and needs to be fundamentally rethought". Hickel's analysis does not confront the data which show strong reductions in stunting over the MDG period (UNICEF/WHO/World Bank Group 2017). Dalby (Chapter 8) argues that transformative change is required to meet the goals, with reduced consumption of fossil fuels in the North. If Weber (Chapter 14) is correct and there is a major shift towards socially responsible investment and/or social investment (as defined in that chapter), then the investment modalities discussed below may be less necessary.

The evolution of development finance

The large majority of low- and middle-income countries (LMICs) were independent by the early 1960s, a decade that can be seen as the heyday of ODA. The newly founded Organization for Economic Co-operation and Development (OECD) and its Development Assistance Committee sought to replicate what were perceived as the successes of the Marshall Plan in European reconstruction to support the development of LMICs. In 1970 OECD members pledged to commit 0.7 per cent of national income to ODA, yet over the subsequent (almost) half century, only a handful of countries met this goal. The oil price rises of the 1970s, however, meant that many LMICs were able to borrow cheaply to finance their development, although this was quickly followed by the "Lost Decade for Development" of the 1980s (Singer 1989) when countries struggled to service their debts. The debt overhang persisted into the 1990s, fuelling calls for a Jubilee debt forgiveness at the start of the new Millennium. Although wholesale debt forgiveness for the LMICs did not occur, some of this energy may have contributed to the push for the MDGs.

In order to help meet the MDGs, new organizations coinciding with specific goals were set up, namely Gavi in 2000 (associated with Goal 4 on child mortality) and the Global Fund in 2002 (associated with Goal 6 on combatting HIV/AIDS, malaria and other diseases). The emergence of these new organizations meant that the World Health Organization, the key UN health agency, was to be a technical rather than an implementing agency for the MDGs.

Making progress on the MDGs required financial support from the participant countries, but also needed external financing, particularly for the low-income countries. Planning exercises were undertaken to estimate the cost of reaching the MDG targets, beginning with estimates of the aggregate amount needed to assist the low- and lower-middle income countries to achieve all the goals (UN Millennium Project 2005). The amounts proposed by the UN Millennium Project group (an additional $73bn in 2006, rising to an additional $135bn in 2015), while large, were considered feasible if official development assistance (ODA) were to increase to the familiar 0.7 per cent of national income target.

Subsequently more detailed plans were developed for individual goals and targets – the HIV-AIDS target (Schwartländer *et al.* 2011), maternal, newborn and child health (Maternal, Newborn and Child Health Network for Asia and the Pacific 2009) and nutrition (Horton *et al.* 2010) – which continued to be updated and refined (Bhutta *et al.* 2013; Stenberg *et al.* 2014). The financing goals at the time seemed audacious but, in the health area at least, unprecedented sums were mobilized. Annual development assistance for health, (as reported by the Institute for Health Metrics and Evaluation 2017) which consists of ODA and contributions from foundations, rose from around $11bn in 2000, to $38bn at its peak in 2013. The spending plans also set a tradition of comparing the costs and benefits of the investments.

The SDGs are considerably more ambitious than the MDGs, with 17 SDGs (as compared to eight MDGs), and 169 targets (instead of 27). Whereas the MDGs focused on the LMICs, the SDGs are "one world" goals, meant to apply to all countries. The Secretary-General of the UN initially estimated that to achieve the SDGs would require an annual commitment of $2.5 trillion (UNCTAD 2014) in addition to funds mobilized domestically by LMICs. This was subsequently revised to $6 trillion annually for a ten-year period (Leone 2017). Clearly, mobilizing such funds will require new sources of financing, especially since foreign assistance from countries reporting to the Development Assistance Committee (DAC) of the OECD is currently not increasing.

To some extent the gap can be filled by the new, non-DAC, development donors, in particular China (Wolf, Wang, and Warner 2013) who are providing considerable amounts of foreign assistance. South–South development assistance is estimated at around $20bn (UNDP 2018). Because of the volume of aid that China provides, it ranks in the top-ten donors for health in sub-Saharan Africa (Grépin *et al.* 2014). However even with South–South development assistance, this will not meet the requirements for trillions for the SDGs.

Can "innovative finance" help fill the SDG funding gap? Some forms of innovative finance for the MDGs raised substantial sums, such as the airline ticket tax whose proceeds go largely to priority infectious diseases through Unitaid, which had raised $1bn prior to 2015 (Stevenson and Whiteside 2015). Efforts to emulate this have been less successful, however. One costly endeavour, Massivegood, launched with the backing of Bill Clinton, Gordon Brown, and Ban Ki Moon, with celebrity endorsements from several well-known actors and sports stars, offered consumers in the travel industry the opportunity to make a microdonation when booking their travel. The scheme was allotted $11m in seed funding, but was discontinued after 16 months, having raised only about $300,000 (May 2011). Other schemes, such as Project Red and Comic Relief, continue to raise funding for health. We will not discuss these mechanisms further here since they are well covered in the existing literature (Connor, Evans, and Brink 2011).

To put those trillions in context, as of 2015, the big five international development banks had about $500bn outstanding in loans and China had $838bn outstanding in loans and deposits to LMICs, while the stock of foreign direct investment to LMICs was $778bn – just over half of all foreign direct investment (Runde 2015). The MSCI ACWI Investable Market Index (which covers 99 per cent of publicly traded equities globally) amounts to $66tn while the volume of investment-grade bonds stands at 51tn (Norman Cumming, personal communication, 30 April 2018). World GDP in 2016 was around $86tn (World Bank 2018), so $6tn represents 7 per cent of world GDP. At the same time, there has been a rising interest among the Millennial generation and the tech and financial communities (among others) in aligning their philanthropy and investments with the SDGs. Climate change is the biggest concern for this group according to the 2017 Global Shapers Survey (and the 2015 and 2016 surveys) undertaken for the World Economic Forum (World Economic Forum 2017).

One general tendency common to all the newer modalities has been a focus on value-for-money. In part, this may have come from the frustrations experienced with foreign aid, where a very large literature debates whether it has been effective and/or how to make it more so. One of the consequences has been a shift towards more public–private partnerships, a focus on metrics, and various forms of management-by-results. At the same time, there is a delicate balance between involving the private sector and "leaving no-one behind", which typically requires oversight and intervention by the public sector.

With a focus on the implications for health, the next section discusses philanthropy from private individuals and foundations, and this is followed by a section discussing loans (blended finance) and another section discussing impact investments and other social investments. This is a fast-moving field and much of the discussion relies on websites and a few interviews rather than analysis of the scholarly literature.

Private philanthropy

Private donations are likely understated as data sources are not complete and tend to be underreported in countries with less comprehensive tracking systems. US citizens and organizations contribute the largest recorded amount of private philanthropic flows internationally (Development Initiatives 2016), with the largest share from the US coming from individuals. Private charitable donations in 2016 from the US were more than five times larger than those of foundations and 16 times larger than those of corporations (many corporations undertake their philanthropic initiatives via separate foundations: The Giving Institute 2017). In 2016, charitable donations in the US amounted to $390bn, although the categories "health" and "international affairs" receive only a modest share of philanthropy as compared to religion and education which together accounted for almost half (although some of the funding attributed to religion may be used for health purposes).

Private charitable donations for international development in 2015 were more than a quarter as large as ODA (Development Initiatives 2016), and all private flows (remittances plus charitable donations) exceed ODA. Private flows are particularly important for rapid-onset emergencies and natural disasters (Development Initiatives 2016) and new modalities such as texting small amounts in response to humanitarian emergencies have made it easier to direct money internationally. Some examples of giving directed to international health are shown in Table 13.1.

Private foundations contribute significant amounts to international development, of which US foundations account for 92 per cent of the reported global total (Development Initiative 2016). The current phase of globalization bears some similarities to the previous major phase (from the 1870s to the eve of World War I), which was associated with rising inequality and the establishment of new foundations with social concerns such as the Rowntree (established 1904), Carnegie (1905) and Rockefeller (1913) Foundations. In the current era, the astronomical increase in number and wealth holdings of the world's billionaires has fuelled the emergence of new philanthropy and new foundations, many from individuals from the tech and financial sectors. A couple of examples focusing on health, both founded in 2002, include the Children's Investment Fund Foundation (CIFF) and the Clinton Health Access Initiative (CHAI).

The Gates Foundation, which is the world's largest philanthropist (IHME 2017), contributed more international assistance than Denmark in 2013, and almost as much as Switzerland (Development Initiatives 2016). In the health field, it is an even bigger player. In 2016, the Gates Foundation contributed $3.3bn for global health, an amount equal to what the UK contributed to health and more than any other single bilateral source except the US (IHME 2017). Bill and Melinda Gates also helped found The Giving Pledge in 2010, a group of 175 billionaires who have pledged to give away over 50 per cent of their wealth either over their lifetimes or in bequests (see Table 13.1). They have encouraged a couple of the other members of the Forbes' list of the top-ten

Table 13.1 Examples of philanthropy related to global health

Organization	Source of funds	Mechanism	Health investment examples
Giving what we can www.givingwhat wecan.org (UK)	Private individuals	Philanthropy of members (pledge >10 per cent of income; recommendations to others	Recommends top charities researched by Givewell
Givewell (USA) www.Givewell.org	Private individuals	Began as recommendations for 8 founder members; now also acts as a charity navigator	Recommend Effective Altruism funds; top-nine charities recommended include seven in health
The Giving Pledge www.Givingpledge. org	High net worth individuals (billionaires)	Pledge to give 50 per cent or more of assets over lifetime/as a bequest; 175 members in 22 countries: created 2010	Individuals donate according to own interests; but several focus on global health

wealthiest individuals/couples in the world (Warren Buffet, Mark Zuckerberg and Priscilla Chan) to focus their philanthropy on health.

The Gates Foundation's influence on global health extends beyond funding. The Foundation has had an important influence through the support it gave to the Global Fund for AIDS, TB and Malaria, and Gavi; indeed, the setting-up of Gavi was catalysed by a challenge Bill Gates issued to key stakeholders at a dinner at his home in 1998 (Gavi 2018). Gavi and the Global Fund have adopted results-oriented approaches, listing numbers of lives saved prominently on their websites, and using mechanisms to "shape" markets, such as making advance guarantees of large volume purchases to help drive down prices for antiretroviral medications for HIV-AIDS, artemisinin combination therapies for malaria, rapid diagnostic tests for TB, and priority childhood immunizations. The Gates Foundation continues to help set up new organizations such as the Bill and Melinda Gates Medical Research Institute and is a major partner in other public–private partnerships such as the new Coalition for Epidemic Pre- paredness Innovations. The Gates Foundation is also a prominent funder of articles published in *The Lancet*, among the highest impact-factor journals cur- rently published. The Gates Foundation has emphasized metrics (obviously key, if one wants to measure impact), both on the burden of disease and on cost- effectiveness of interventions.

The emphasis on metrics and impact has been a broader trend across private philanthropy. "Charity navigators" which provide advice on which charities make the best use of funds have become more prominent since 2000. The US- based Charity Navigator (charitynavigator.org) was founded in 2001 by a phar- maceutical executive and philanthropist and provides statistics to inform donors (both domestically and internationally). Among their eight "hot topics" in April 2018, one focuses on the SDGs, along with a list of charities contributing to each of the 17 SDGs, with a star rating of their effectiveness (Charity Navig- ator 2018). GiveWell, part of a larger movement known as "Effective Altruism", originated with a group of eight friends working in the financial industry who wanted to ensure that their donations to international development were effective (GiveWell 2018a). GiveWell recommends a small group of charities that it ranks as effective and has directed $60m since 2007 to its list of top char- ities (Pitney 2017). As of April 2018, seven of the top nine focus on global health (GiveWell 2018b). Those using GiveWell's recommendations include the Facebook co-founder Dustin Moskovitz (Pitney 2017) and another giving "club", Giving What We Can founded in 2009 and based in the UK, where members commit to giving 10 per cent or more of their income over their life- time to effective charities.

Private individuals and foundations do not only provide funding as gifts: they also direct their investments in socially conscious ways, through impact invest- ments, which are discussed later below, following section on blended finance (blending public and private lending).

Development impact bonds and other forms of blended finance

Development Impact Bonds (DIBs) are a variant of Social Impact Bonds. The UK arranged the first Social Impact Bond in 2010, since adopted in the US, Canada, Australia, and several other European countries. As of the first of August 2017, nearly 90 DIBs had been contracted (Boggild-Jones and Gustafsson-Wright 2017). Social Impact Bonds are utilized to tackle important social issues with measurable outcomes, in such a way as to incentivize effectiveness of service providers. Private funders provide the initial financing for service providers to work towards agreed-upon outcomes. If the goals are achieved in the specified time frame, public funders (domestic governments) then repay the initial investors. This requires appropriate legal and regulatory conditions, and the expertise of at least two other groups: an intermediary which undertakes the technical work to design the bond and an independent evaluator who determines whether the outcomes were achieved. The mechanisms and success of Social Impact Bonds have been reviewed elsewhere (Gustafsson-Wright, Gardiner, and Putcha 2016). The advantages include the assumption of some of the risk by the private sector who provide funds at relatively favourable rates, and the potential encouragement of service providers to be innovative and effective in delivering services. The bonds also allow private investors to feel good about supporting a social cause, while still earning returns on their investment.

DIBs are a more recent variant, where international funding agencies replace domestic governments as the ultimate funder, and the projects involved are in international development. These are still a relatively new phenomenon. As of the first of August 2017, only four had been contracted, although many more are currently in development (Boggild-Jones and Gustafsson-Wright 2017). Table 13.2 summarizes information on two existing DIBs in health and two others in development. Of 24 DIBs in development as of August 2017, the largest number (11) were in health (Boggild-Jones and Gustafsson-Wright 2017).

Certain areas of health lend themselves to DIBs. One of the four DIBs discussed in Table 13.2 and one of the blended finance interventions in Table 13.3, benefit in eye hospitals (primarily cataract operations). Cataract operations are a health intervention known to have high cost-effectiveness (Lansingh, Carter, and Martens 2007) and where there has been a large backlog of those desiring treatment at an affordable cost. There is also potential for the private sector to provide these services sustainably while still making them accessible to those with low income, using cross-subsidization. The Cameroon DIB (Table 13.2) provides a bonus payment to the service provider, if 40 per cent of the services go to households in the bottom two income quintiles (Oroxom, Glassman, and McDonald 2018). There is potentially a broad scope for using the DIB mechanism to upgrade accreditation of health service providers, as in the Rajasthan DIB (Table 13.2), which focuses on private facilities providing maternal and newborn care. Problems of low quality of services are endemic in South Asia and sub-Saharan Africa and improving the proportion of accredited providers offers a lot of scope.

Table 13.2 Development impact bonds (DIBs) related to global health

Organization	Source of funds	Mechanism	Health investment examples
Rajasthan MCN health DIB[1]	Private foundations; USAID	Involves a design agency and an implementing agency	Improving quality and attaining accreditation standards for 440 private healthcare facilities
Cameroon Optical Care DIB[2]	Private foundations; OPIC	Same as above	Provide up to 18,000 cataract surgeries over five years, to reduce major backlog
Cameroon Kangaroo Mother Care DIB[3] (in design)	World Bank, International NGOs, bilateral donors, Cameroon Ministry of Health	Same as above	Implement Kangaroo Mother Care in up to 12 public hospitals, to save lives of premature and low-weight newborns
Mozambique Malaria Performance Bond[4] (pilot mode only)	Private foundations, Mozambique Ministry of Health, others	Same as above	Aimed to reach 8m people over 12 years

Notes
1 www.devex.com/news/usaid-announces-a-new-development-impact-bond-91621.
2 www.opic.gov/press-releases/2017/cameroon-cataract-development-impact-loan-offers-innovative-approach-prevent-blindness.
3 www.socialfinance.org.uk/projects/cameroon-kangaroo-mother-care.
4 www.devex.com/news/have-development-impact-bonds-moved-beyond-the-hype-88372.

The outcomes of these early DIBs will undoubtedly be watched with interest. At the moment, because the modality is still new, bonds are difficult and time-consuming to develop, and regulatory frameworks are nascent. Impact Investment Shujog Limited (2014) strikes a note of caution, noting that for every ten DIBs discussed, work is done on three or four, and one goes through. Boggild-Jones and Gustafsson-Wright (2017) argue that establishing funds (with several member projects) rather than individual bonds is a way to scale up, an approach which has also been used for Social Impact Bonds.

DIBs are one form of blended finance, and there are a variety of other arrangements which blend public and private funding: Table 13.3 provides seven different examples from the health area, five of which involve the World Bank. The multilateral development banks (MDBs) arguably have a key role in helping to mobilize private capital to support the SDGs. Jim Yong Kim, former President of the World Bank, wanted to increase the current $7bn from the private sector, to $30bn (Thomas 2018). To do so he had to steer between those on the right who think the MDBs should mostly be abolished (discussed by Morris 2018) and those on the left concerned he spent too much time rubbing shoulders with the wealthy at Davos.

Table 13.3 Examples of blended finance

Organization	Source of funds	Mechanism	Health investment examples
Global Health Innovation Fund www.ghif.com	Two foundations, two bilateral donors, four banks, three drug companies, IFC	$108m social impact fund for late-stage development of vaccines, drugs, diagnostics and other interventions for diseases in LMICs; project must become commercially viable.	Chikunyunga vaccine; oral cholera vaccine; TB diagnostics; diagnostics for pre-eclampsia
IFFIm (Innovative Finance Facility for Immunisation) www.iffim.org	9 donor countries (a tenth one has pledged to join) back bonds held by private sector	World Bank issues Gavi bonds backed by $5.7bn in donor commitments, to purchase vaccines immediately and get volume discounts; repaid from donor commitments	Increased demand for pentavalent vaccine and reduced cost for set of essential vaccines; delivered $2.5bn to support vaccines
Convergence Finance www.convergence.finance	Two foundations, one government, private finance	Uses core funding to design complex bonds to leverage private sector funding (see above: provided design funding for Rajasthan Development Impact Bond)	Alina Vision received $12m in equity and grant funding for 60 affordable eyecare hospitals;

Pandemic Emergency Financing Facility http://treasury.worldbank.org/cmd/htm/World-Bank-Launches-First-Ever-Pandemic-Bonds-to-Support-500-Million-Pandemic-Emergenc.html	Two bilateral donors	Bonds and derivatives	$425m was raised on bond market, to provide $500m in insurance against six diseases, with cash to help with outbreaks of selected other diseases; for IDA-eligible countries
Power of Nutrition www.powerofnutrition.org/who-we-are/ began 2015	World Bank, two foundations, bilateral donors, UNICEF	Guarantees that investments will receive a 4x match; uses payment by results mechanism	Interventions in nutrition supplements, education and services in five "hotspot" countries
Health Credit Fund www.opic.gov/press-releases/2016/medical-credit-fund-raises-additional-usd-17-million-healthcare-impact-investment-africa	OPIC, two bilateral donors, two foundations, private lenders	Provides loans for healthcare providers in Africa; partner Pharmaccess provides training to improve quality standards via SafeCare initiative	$16m in loans disbursed as of 2016 in five countries; continuing to expand
Global Financing Facility www.globalfinancingfacility.org	Bilateral donors fund trust fund; World Bank loans; domestic finance and private finance	Funds country health priorities	Raised $800m initially, now seeking replenishment to $2bn

The seven health interventions in Table 13.3 bring in different amounts of private funding, ranging from the low tens of millions (Medical Credit Fund; Convergence Finance's Alma Vision project), to hundreds of millions (Global Health Innovation Fund; Pandemic Financing Facility; Power of Nutrition) to one or two billion (International Finance Facility for Immunization, Global Financing Facility).

Impact investment and investments directed at social good

Investments which take into account ethical considerations can be described in various ways. One source (Zhou 2018) defines three different variants. First, investments may be rated using Environmental, Social and Governance (ESG) scores, related to the "double bottom line" term coined in the 1990s and subsequently expanded into the "triple bottom line". Second, socially responsible investments are investments where a screen has been used, to screen out investments which do not accord with the investor's ethical choices (see Weber Chapter 14). Socially responsible investments are a smaller group than all investments with good ESG scores. Third, impact investments are those that have positive effects on some social goal.

The term "impact investment" was first coined in 2007 as a result of a meeting convened by the Rockefeller Foundation (Kanani 2012), although the ideas had already been around since at least the early 1990s. In 1997 the Heron Foundation moved to invest a proportion of its endowment portfolio into "Mission-Related Investments", deciding that "the foundation should be more than a private investment company that uses its excess cash flow for charitable purposes" (Heron Foundation 2018). Impact investment is related to the "double bottom line" and to social entrepreneurship. A timeline of some of the key steps in the development of impact investing can be found in Toniic (2016).

Impact investing to support the SDGs is sufficiently widespread that the UN has issued principles for "positive finance" (UNEP 2018), and the G8 issued a report from its Social Impact Investment Taskforce (Social Impact Investment Taskforce 2014). It has expanded rapidly and there are various modalities such as peer groups of wealthy individuals/families (for example the Pymwymic Group in Europe, and Investor's Circle in the US: see Table 13.4), investment companies catering to high net-worth individuals with social goals (e.g. Tribe Impact Capital in the UK), and angel investment peer groups (Clearly Social Angels, UK). ETHEX (a small not-for-profit in the UK) offers opportunities for retail investors to help fund social enterprises or to provide long-term bonds to charitable organizations. GIIN, TONIIC and Sonen Capital undertake investments for individuals, foundations and organizations. What these various modalities have in common is that they typically fund investments in the areas of the environment, health, education, and other social areas. Some examples of impact investments in global health are listed in Table 13.4.

The Netherlands has been a leader in aligning investments with the SDGs. In 2016, 18 Dutch financial institutions (including banks, insurance companies and

Table 13.4 Examples of impact investment

Organization	Source of funds	Mechanism	Health investment examples
Individual investors			
Pymwymic Group, (Europe[1]) www.pymwymic.com	Private wealth	Funds early-stage for-profit companies with social impact, via direct investments and through funds. Also donate funding	n/a
Clearly Social Angels (UK) www.clearlyso.com	Private wealth valued at £175m	Angel investment, debt or equity	https://newatlas.com/eyejusters-adjustable-glasses-developing-world/22734/ (other examples for high-income countries)
Investor's Circle (US) www.investorscircle.net	Private wealth	Combines impact investing and social entrepreneurship groups; early stage investments	www.thinkmd.org uses mobile technology to help generate clinical assessments; AYZH clean birth kits www.ayzh.com/#home
Tribe Impact Capital www.tribeimpactcapital.com (UK)	Private wealth	Manage portfolios for clients with specific impact investment goals, in support of SDGs	n/a
Sonen www.sonenCapital.com (US)	Private wealth, foundations, firms	Manages portfolios allied to SDGs for high net worth individuals, and offers funds for individuals, foundations and organizations	n/a
ETHEX www.ethex.org.uk	Retail investors	Small not-for-profit which offers shares in social enterprises, bonds for long-term loans to charitable organizations	Mainly in UK; some global interests, but none listed currently in health

continued

Table 13.4 Continued

Organization	Source of funds	Mechanism	Health investment examples
TONIIC www.toniic.com	Members: individuals, family offices, foundations	Not-for-profit member-driven organization providing transformational funding to social enterprises; members collectively have $1.6bn committed to impact investment (about half of this from investors whose portfolio is 100 per cent impact investment)	Carego International – affordable clinics, self-contained suites with power and communications; Clínicas del Azúcar (diabetes clinics) in Mexico; Healthpoint Services: rural India "tele-doctors" + diagnostic tests
Organizations investing GIIN (Global Impact Investing Network) https://thegiin.org	Represent impact investment (II) organizations not investors, who have either committed $10m to II or made 5 II, or both	60 per cent of investors either already track, or plan to track, investments with respect to SDGs	Healthcare is second largest investment area, after food and agriculture
Big Society Capital Bigsociety capital.com Founded 2008	£600m, 2/3 from Reclaim Fund (dormant accounts) and balance from £50m contributions from each of four major banks	Invest in specialized funds, general funds, social impact bonds and intermediaries	n/a

Note
1 "Put your money where your meaning is"; n/a means not available.

pension funds) launched an SDG investment agenda. They aim to deploy blended finance investments, and to make SDG-aligned investments "the new normal", available to retail investors. The two biggest pension funds (ABP and PGGM) aim to invest €8bn in Sustainable Development Investments by 2020 (PGGM 2018).

Other countries have devised innovative schemes to harness investments for social good. In the UK, the Dormant Bank and Building Societies Accounting Act passed in 2008, allowed £400m in funding from dormant accounts to be used for social investments (while providing for payment in case accounts were claimed), to which four UK banks each added £50m, making a total fund of £600m, to create Big Society Capital. Big Society Capital estimated that in 2016 in the UK there was almost £2,000m in outstanding social capital loans in the UK, of which £142m (7 per cent) was from Big Society Capital itself (Big Society Capital 2017).

France has a long history of solidarity movements and various regulations and tax exemptions encourage all companies with more than 50 employees to offer at least one "solidarity finance scheme" as a savings vehicle for their employees. These schemes invest a portion of their funds in social products, and some donate a portion of the returns to charity. They control substantial amounts of assets (£6.84bn in 2014: Numbers for Good 2016). In some of the Southern European countries, social investments have a somewhat different format. In Italy, public savings banks set up in the nineteenth century engaged in lending, but also were required to undertake social and philanthropic activities. Legal changes in the 1990's split the commercial and philanthropic aspects of the banks, however the foundations and banks remain somewhat intertwined. This history has led to a relatively large number of charitable foundations in Italy (The Economist 2001).

Pension funds in other countries which already consider the SDGs in making their investments include Australia's Cbus Super (the Construction and Building Union, with $2.7bn in funds: Purves 2017), and CalPERS (the California Public Employees Retirement System). CalPERS, whose $357bn pension fund is one of the largest in the US, already considers ESG criteria (environment, sustainability and governance) and is now considering the importance of the SDGs (Diamond 2018). In a recent initiative, a group of pension funds announced following the G7 meeting in Canada in June 2018, that they would collaborate to address the G7 priorities of climate change, gender inequality and the infrastructure gap (Ontario Teachers' Pension Plan 2018). In addition to PGGM and CalPERS (both mentioned here already) these included five major Canadian public sector pension plans.

Social Stock Exchanges are another mechanism for linking investors with companies, The UK Social Stock Exchange raised £400m for member companies in 2015 (Chynoweth 2016). Canada has a similar platform SVX which links investors with private market securities; and the Impact Exchange based in Mauritius lists African and Asian enterprises.

Impact and socially directed investments would seem to be an area for potential growth, particularly if major institutional investors such as pension funds with billions of dollars in assets move in this direction. Pension funds tend to be more conservative and less nimble than individual investors and foundations, but as experience builds up in this sector, more funds may follow the examples of those in the Netherlands, Australia, Canada and California mentioned above. One area in need of development is that of data/metrics. The Impact Management Project (2018) is one organization working to set shared norms which can be used to evaluate portfolios.

This is an evolving area. The Heron Foundation, the first one to move into impact investment for its portfolio, began to move to "100% impact" in 2012, and succeeded in doing so by 2017. Toniic (2016) reported that of its 160 members, 85 were committed to 100 per cent impact. The Ford Foundation committed in 2017 to moving $1bn of its $12bn endowment to "mission-related investments", the largest such commitment by a foundation to date (Ford Foundation 2017). US private foundations in the US own $865bn in assets, so this could be a significant source of socially directed investment.

Conclusions

To succeed in moving "from billions to trillions" in order to support the SDGs, it will be essential to involve the private sector. There will, however, continue to be an essential role for the public sector in terms of ensuring the provision of public goods, and "leaving nobody behind".

Since 2000 there have been rapid changes and innovations in financing possibilities which can potentially support the SDGs. The mindset of the Millennials (and generally growing concern for the environment) have been important factors. Concerns about the environment and about growing inequality have been somewhat of a catalyst for interest in social causes more generally. Another factor is the vast new wealth of those benefiting from the tech revolutions (the FAANG economy, the finance sector), who (due to global capital mobility) face lower tax burdens, and some of whom have set up new foundations with social mandates.

There are issues still needing work. Private philanthropy is obviously important and has the advantage of its greater flexibility. Philanthropists can take more risks with their gifts and hence promote innovative areas, which can then be funded at scale through other modalities. There may also be demonstration effects to the general public and to taxpayers. However, the scale of private philanthropy is smaller than that of the other mechanisms discussed.

Although the DIBs and blended finance instruments show considerable promise, the DIBs in particular are labour-intensive to develop, and only a fraction of those on which initial work is done, are actualized. The initial DIBs are in the $10m vicinity which is appropriate for a new mechanism, but it would take a lot of these to reach a billion (let alone a trillion) dollars. An advantage, however, is that DIBs can be developed to meet specific social challenges which the private equity market will not identify or will not prioritize independently.

Impact and socially directed investment have the greatest potential in terms of volume, given the value of funds concerned (holdings of pension funds, private foundations etc., do add up to trillions of dollars). Mapping exercises for impact investment for pension funds show that it can be difficult to identify investments which truly have a positive impact and/or align with the SDGs. Generating the appropriate information can be time-consuming. Thus, impact investments may only loosely align with the SDGs, and the main investments will likely be in areas with greater commercial potential, generally benefiting the rising middle class rather than the poor, globally.

It is possible that the more targeted approach open to private philanthropy and the blended finance mechanisms, can open up areas for future larger volumes of impact and social investments. The impact investment arena may need to either draw on the expertise of the charity navigators (from the philanthropy space) or the intermediaries (from the DIB/blended finance space) with subject matter as opposed to financial expertise. The UN agencies' and multilateral development banks' work in assembling evidence on benefit–cost or cost-effectiveness of social interventions, has also been essential for identifying potential impact investments. Without studies of the economics of early childhood education from the last three or four decades, and similarly of the economics of nutrition and of health from the last two or three decades, it would have been difficult to identify the current large-scale, impactful interventions. There are areas of the SDGs where work is less far advanced, such as how to intervene to prevent gender-based violence or against children, where more research and study is required.

All these mechanisms are information-intensive. Identifying investments which not only have good financial returns but also good social returns, requires expertise. On the flip side, perhaps this will be a way to harness the expertise of fund managers whose functions are threatened by developments such as exchange-traded funds (ETFs).

The Secretary-General of the UN (UN Secretary General 2015b) called for policy commitments in six crucial areas to support the SDGs, in his remarks at the opening of the Third International Conference on Financing for Development. One of these is particularly relevant to governance, and to the financing of the goals, namely the need for international co-operation in tax matters, in order to reduce illicit flows. This is a particularly thorny area of governance. The World Bank (2017) noted that although there has been action for at least two decades "a new and invigorated approach is needed". This is a topic also of great importance regarding SDG 10 on inequality and is discussed in more depth in Chapter 1 (Dalby, Horton, and Mahon).

At first sight, the sheer number of the SDGs, their one-world focus, and the volume of investment required (7 per cent of world GDP annually, for ten years) seems daunting. However, the consensus generated by the bottom-up goal-setting exercise, and some of the financial evolutions discussed in this chapter, provide some grounds for hope that progress on these ambitious goals is possible.

Acknowledgements

Thanks for useful suggestions and comments to Norman Cumming from CR Global LLP, Karen Grépin from Wilfrid Laurier University and James Lawson from Tribe Impact Capital.

References

Anders, M. 2016. "Why this Nobel Prize-winning Economist Believes the Data Behind the SDGs 'Doesn't Add Up'". *Devex News*, 14 July. Accessed 12 September 2018. www.devex.com/news/why-this-nobel-prize-winning-economist-believes-the-data-behind-the-sdgs-doesn-t-add-up-88417.

Bhutta, Z. A., J. K. Das, A. Rizvi, M. F. Gaffey, N. Walker, S. Horton, P. Webb, *et al.* 2013. "Evidence Based Interventions for Improving Maternal and Child Nutrition: What Can Be Done and at What Cost?". *The Lancet* 382 (9890): 452–477.

Big Society Capital. 2017. "Size of the Social Investment Market". Big Society Capital. Accessed 19 April 2018. http://bigsocietycapital.com/home/about-us/size-social-investment-market.

Boggild-Jones, I. and E. Gustafsson-Wright. 2017. "What's Next for Impact Bonds in Developing Countries?" *Brookings*, 28 September. Accessed 18 April 2018. www.brookings.edu/blog/education-plus-development/2017/09/28/where-next-for-impact-bonds-in-developing-countries/.

Charity Navigator. 2018. "Hot Topics: United Nations Sustainable Development Goals". Charity Navigator. Accessed 19 April 2018. www.charitynavigator.org/index.cfm?bay=content.view&cpid=4742.

Chynoweth, C. 2016. "Alternative Stock Exchange Promotes Both Profit and Social Impact". *Guardian*, 27 May. Accessed 22 September 2018. www.theguardian.com/sustainable-business/2016/may/27/alternative-stock-exchange-promotes-profit-and-social-impact.

Connor, D., D. Evans and B. Brink. 2011. "Private Sector: New Ways of Doing Business". In *Innovative Health Partnerships: The Diplomacy of Diversity*, edited by D. Low-Beer, 333–348. Singapore: World Scientific Books.

Development Initiatives. 2016. "Private Development Assistance Key Facts and Global Estimates: Factsheet". Development Initiatives. Accessed 18 April 2018. http://devinit.org/post/private-development-assistance-key-facts-and-global-estimates/.

Diamond, R. 2018. "CalPERS Examines Adopting SDGs". *Top1000funds.com*, 18 January. Accessed 12 September 2018. www.top1000funds.com/news/2018/01/18/calpers-examines-adopting-sdgs/.

Easterly, W. 2015. "The SDGs Should Stand for Senseless, Dreamy, Garbled". *Foreign Policy*, 28 September. Accessed 12 September 2018. https://foreignpolicy.com/2015/09/28/the-sdgs-are-utopian-and-worthless-mdgs-development-rise-of-the-rest/.

Ford Foundation. 2017. "Ford Foundation Commits $1 Billion from Endowment to Mission-Related Investments". *Ford Foundation*, 5 April. Accessed 20 April 2018. www.fordfoundation.org/the-latest/news/ford-foundation-commits-1-billion-from-endowment-to-mission-related-investments/.

Gavi. 2018. "The History of Gavi". Gavi: The Vaccine Alliance. Accessed 12 September 2018. www.gavi.org/about/mission/history/.

GiveWell. 2018a. "Our Story". GiveWell. Accessed 21 January 2018. www.givewell.org/about/story.

GiveWell. 2018b. "Top Charities". GiveWell. Accessed 19 April 2018. www.givewell. org/charities/top-charities.

Grépin, K. A., V. Y. Fan, G. C. Sen, and L. Chen. 2014. "China's Role as a Global Health Donor in Africa: What Can We Learn from Studying under Reported Resource Flows?". *Globalization and Health* 10 (84): 1–11.

Gustafsson-Wright, E., S. Gardiner, and V. Putcha. 2016. "The Potential and Limitations of Impact Bonds Lessons from the First Five Years of Experience Worldwide". *Brookings*, 9 July. Accessed 19 April 2018. www.brookings.edu/research/the-potential-and-limitations-of-impact-bonds-lessons-from-the-first-five-years-of-experience-worldwide/.

Heron Foundation. 2018. "The evolution of Heron". Heron. Accessed 19 April 2018. www.heron.org/enterprise.

Hickel, J. 2016. "The True Extent of Global Poverty and Hunger: Questioning the Good News Narrative of the Millennium Development Goals". *Third World Quarterly* 37 (5): 749–767.

Horton, S., M. Shekar, C. McDonald, and A. Mahal. 2010. *Scaling-up Nutrition: What Will It Cost?* The World Bank Directions in Development Report, Washington, DC. Accessed 30 September 2018. http://siteresources.worldbank.org/HEALTH NUTRITIONANDPOPULATION/Resources/Peer-Reviewed-Publications/Scaling UpNutrition.pdf.

Impact Investment Shujog Limited. 2014. "Financing Healthcare Services for the Poor". IIX. Accessed 15 April 2018. https://iixglobal.com/download/financing-healthcare-services-poor-2014/.

Institute for Health Metrics and Evaluation (IHME). 2017. *Financing Global Health 2016: Development Assistance, Public and Private Health Spending for the Pursuit of Universal Health Coverage.* IHME Report, Seattle, WA. Accessed 30 September 2018. www. healthdata.org/sites/default/files/files/policy_report/2017/IHME_FGH_2016_ Technical-Report.pdf.

Kanani, R. 2012. "The State and Future of Impact Investing". *Forbes*, 23 February. Accessed 19 April 2018. www.forbes.com/sites/rahimkanani/2012/02/23/the-state-and-future-of-impact-investing/#12ff1283ed48.

Lansingh, V. C., M. J. Carter, and M. Martens. 2007. "Global Cost-Effectiveness of Cataract Surgery". *Ophthalmology* 114 (9): 1670–1678.

Leone, F. 2017. "UNGA Launches Global Consultation on Financing SDGs". International Institute for Sustainable Development (IISD), 20 April. Accessed 5 January 2018. http://sdg.iisd.org/news/unga-launches-global-conversation-on-financing-sdgs/.

Maternal, Newborn and Child Health Network for Asia and the Pacific. 2009. *Investing in Maternal, Newborn and Child Health: The Case for Asia and The Pacific.* Report. World Health Organization and The Partnership for Maternal, Newborn & Child Health. Geneva. Accessed 30 September 2018. www.who.int/pmnch/topics/investing inhealth.pdf.

May, K. 2011. "Massive Good Charity Project Axed, Travel Technology Worked but Brand Failed". *Tnooz Newsletter*, 25 November. Accessed 17 April 2018. www.tnooz.com/article/ massivegood-charity-project-axed-travel-technology-worked-but-brand-failed/.

Morris, S. 2018. "John Bolton Wants to Shut Down the World Bank". *Center for Global Development*, 23 March. Accessed 19 April 2018. www.cgdev.org/blog/john-bolton-wants-shut-down-world-bank.

Numbers for Good. 2016. "France: The Country with One Million Social Investors". *Numbers for Good*, 21 January. Accessed 19 April 2018. http://numbersforgood. com/2016/01/france-the-country-with-one-million-social-investors/.

Ontario Teachers' Pension Plan. 2018. "Leading Canadian and G7 Investors Come Together in Support of Global Development Initiatives". *Ontario Teachers' Pension Plan*, 6 January. Accessed 12 September 2018. www.otpp.com/news/article/-/article/790381.

Oroxom, R., A. Glassman, and L. McDonald. 2018. "Structuring and Funding Development Impact Bonds for Health: Nine Lessons from Cameroon and Beyond". *Center for Global Development*, Policy Paper no. 117. Washington, DC. Accessed 30 April 2018. www.cgdev.org/publication/structuring-funding-development-impact-bonds-for-health-nine-lessons.

PGGM. 2018. "Institutional Investment into The Sustainable Development Goals". Accessed 19 April 2018. www.pggm.nl/wie-zijn-we/pers/Documents/Institutional-investment-into-the-Sustainable-Development-Goals-statement.pdf.

Pitney, N. 2017. "That Time A Hedge Funder Quit His Job and Then Raised $60 Million For Charity". *Huffington Post*, 6 December. Accessed 19 April 2018. www.huffingtonpost.ca/entry/elie-hassenfeld-givewell_n_6927320.

Purves, D. 2017. "Cbus Super to integrate UN's Sustainable Development Goals". *Investment Magazine*, 27 February. Accessed 19 April 2018. https://investmentmagazine.com.au/2017/02/cbus-super-to-integrate-uns-sustainable-development-goals/.

Runde, D. 2015. "Ensuring the World Bank's Relevance". *Forbes*, 16 March. Accessed 10 January 2018. www.forbes.com/sites/danielrunde/2015/03/16/ensuring-world-bank-relevance/#59b1e7597feb.

Schwartländer, B., J. Stover, T. Hallett, R. Atun, C. Avila, E. Gouws, M. Bartos, *et al.* 2011. "Towards an Improved Investment Approach for an Effective Response To HIV/AIDS". *Lancet* 377 (9782): 2031–2041.

Singer, H. W. 1989. "The 1980s: A Lost Decade – Development in Reverse?". In *Growth and External Debt Management*, edited by H. W. Singer and S. Sharma, 46–56. London: Palgrave Macmillan.

Social Impact Investment Taskforce. 2014. "Impact Investment: The invisible Heart of Markets: Harnessing the Power of Entrepreneurship, Innovation and Capital for Public Good". Accessed 7 October 2018. https://impactinvestingaustralia.com/wp-content/uploads/Social-Impact-Investment-Taskforce-Report-FINAL.pdf.

Stenberg, K., H. Axelson, P. Sheehan, I. Anderson, M. Gülmezoglu, M. Temmerman, E. Mason, *et al.* 2014. "Advancing Social and Economic Development by Investing in Women's and Children's Health: A New Global Investment Framework". *Lancet* 383 (9925): 1333–1354.

Stevenson, M., and A. Whiteside. 2015. "Innovative and Ethical Financing for Health for Marginalized Populations". Draft to The Global Fund and the African Development Bank, Balsillie School of International Affairs.

The Economist. 2001. "Italy's Charitable Foundations: Odd Sort of Ownership". *The Economist*, 25 October. Accessed 19 April 2018. www.economist.com/node/835234.

The Giving Institute. 2017. *"Giving USA 2017"*. Giving USA, 12 June. Accessed 17 April 2018. https://givingusa.org/tag/giving-usa-2017/.

The Impact Management Project. 2018. "About". https://impactmanagementproject.com/about/. Accessed 7 October 2018.

Thomas, L. 2018. "The World Bank Is Remaking Itself as a Creature of Wall Street". *New York Times*, 25 January. Accessed 21 April 2018. www.nytimes.com/2018/01/25/business/world-bank-jim-yong-kim.html.

Toniic. 2016. *Launch Report: T100*. Toniic. Accessed 15 April 2018. www.toniic.com/launch-report-download/.

UN Millennium Project. 2005. *Investing in Development: A Practical Plan to Achieve the Millennium Development Goals*. New York. Accessed 30 September 2018. http://site resources.worldbank.org/INTTSR/Resources/MainReportComplete-lowres%5B1 %5D.pdf.

UN Secretary General. 2015a. "Secretary-General's Remarks at World Bank Event "Billions to Trillions – Ideas to Action". *United Nations Secretary-General*, 13 July. Accessed 11 September 2018. www.un.org/sg/en/content/sg/statement/2015-07-13/ secretary-general%E2%80%99s-remarks-world-bank-event-%E2%80%9Cbillions-trillions-%E2%80%93.

UN Secretary General. 2015b. "Secretary-General's Remarks at Opening of Third International Conference on Financing for Development". *United Nations Secretary-General*, 13 July. Accessed 11 September 2018. www.un.org/sg/en/content/sg/statement/ 2015-07-13/secretary-general%E2%80%99s-remarks-opening-third-international-conference.

UNCTAD (United Nations Conference on Trade and Development). 2014. *World Investment Report 2014: Investing in the SDGs: An Action Plan*. Accessed 8 January 2018. http://unctad.org/en/PublicationsLibrary/wir2014_en.pdf.

UNDP (United Nations Development Programme). 2018. *Financing The 2030 Agenda: An Introductory Guidebook for UNDP Country Offices*. Accessed 12 September 2018. www.undp.org/content/undp/en/home/librarypage/poverty-reduction/2030-agenda/ financing-the-2030-agenda.html.

UNEP (United Nations Environment Programme). 2018. *The Principles for Positive Impact Finance: A Common Framework to Finance the Sustainable Development Goals*. Geneva: United Nations Environment Program Finance Initiative. Accessed 19 April 2018. www.unepfi.org/positive-impact/positive-impact/.

UNICEF/WHO/World Bank Group. 2017. *Global Database on Child Growth and Malnutrition: Joint Child Malnutrition Estimates – Levels and Trends*. Accessed 12 September 2018. www.who.int/nutgrowthdb/estimates2016/en/.

Wolf, C., X. Wang, and E. Warner. 2013. *China's Foreign Aid and Government-Sponsored Investment Activities: Scale, Content, Destinations, and Implications*. RAND Corporation, Santa Monica CA. Accessed 12 September 2018. www.rand.org/pubs/research_reports/RR118.html.

World Bank. 2017. "Illicit Financial Flows (IFFs)". *The World Bank*, 7 July. Accessed 12 September 2018. www.worldbank.org/en/topic/financialsector/brief/illicit-financial-flows-iffs.

World Bank. 2018. "World Bank Open Data". World Bank. Accessed 30 April 2018. http://data.worldbank.org.

World Economic Forum. 2017. *Global Shapers Survey*. Annual Survey 2017. Accessed 30 April 2018. http://shaperssurvey2017.org/static/data/WEF_GSC_Annual_Survey_ 2017.pdf.

Zhou, M. 2018. "ESG, SRI & Impact Investing: Explaining the Difference to Clients". *Investopedia*, 2 January. Accessed 12 September 2018. www.investopedia.com/financial-advisor/esg-sri-impact-investing-explaining-difference-clients/.

14 Sustainable finance and the SDGs

The role of the banking sector

Olaf Weber

Finance is a keystone in the transition to sustainable development
(Zadek and Robins 2018, 36)

Introduction

With products such as socially responsible investment, environmental and carbon finance, micro-finance, or green bonds, the financial sector seeks to contribute positively to sustainable development, including the Sustainable Development Goals (SDGs). The financial sector, however, also contributes to military interventions, growing disparity of incomes, de-coupling of finance and real economy, persistence of unsustainable industrial practices that are responsible for environmental degradation and climate change as well as for irresponsible labour practices, and global economic crises. Of course, there is a big variance in the banking sector about how to address sustainability. Microfinance organizations directly address SDG 1 "No Poverty", climate bonds address SDG 13 "Climate Action", many credit unions' business address SDG 11 i.e. to finance sustainable cities and communities, and many global banks address SDG 5 "Gender Equality" through their employment policies. Royal Bank of Canada (RBC), for instance, addresses closing the gender gap since 1994 as one its main priorities in corporate social responsibility.[1]

With the exception of financial institutions that directly address sustainability issues, the integration of sustainability issues into finance decisions is mainly driven by risk management purposes, new business opportunities, or cost savings (Weber and Feltmate 2016). Still, banks that announce that they will dedicate a part of their business to sustainable development, such as low carbon finance or education, are lauded. However, addressing sustainable development should be part of their core business strategy if they take sustainable banking seriously and try to differentiate sustainable banking from socially responsible investing (SRI) and impact investing. Sustainable banking could be defined as a form of banking that addresses sustainable development, including the SDGs, as a core business strategy and not just as a niche-product. SRI mainly integrates environmental, social, and governance criteria into financial decision-making without having the explicit goal to address sustainable development. Furthermore, impact investing often picks a certain topic that is related to the environment and societal needs to invest in. Hence, sustainable development could be such a topic, but it is not the only one that is addressed by impact investing.

Still, the financial industry's motivations to pursue sustainable practices are primarily reactive rather than proactive. Furthermore, the role of the financial sector is ambivalent as it contributes, sometimes even at the same time, both to the causes of sustainability problems as well as to possible solutions to these problems through a variety of different investment and lending practices (Wiek and Weber 2014). The same lender, for instance, might finance green energy projects and polluting coal power plants at the same time.

In addition, links that are more indirect exist between finance and the SDGs. Although processed food is not the only cause of childhood obesity, through financing global processed food producers, banks might negatively contribute to childhood obesity and hence work against SDG 3 "Good Health and Well-Being" (Moodie *et al.* 2013; Monteiro *et al.* 2013). Nevertheless, financiers should be aware that they might be involved in activities that contradict sustainable development and that they might have the opportunity to influence their clients to address these controversies. For instance, even seemingly neutral infrastructure finance that prefers car use to walking and cycling might negatively contribute to SDG 3.

One reason for this ambivalent role of the banking sector with regard to sustainable development is due to its "business case of sustainability" (Schaltegger and Burritt 2015) approach. This approach favours addressing sustainability issues as long as they are material for the bottom line. Hence, only issues that have an immediate and direct impact on business risks and opportunities are considered valuable. In contrast to this approach, other strategies go beyond the business case of sustainability (Dyllick and Hockerts 2002) and apply the sustainability case of business – or in our case the sustainability case of banking and finance (Weber 2014; Weber and Feltmate 2016).

The difference is that the sustainability case of finance identifies the main sustainability issues first, and then tries to create business approaches to address the issue. Financial institutions, for instance, could analyse the 17 SDGs and find lending, investment, and asset management products and services addressing these goals. Furthermore, they could analyse all the goals with regard to the degree to which financial products and services have an effect on them.

The SDGs are couched in terms of transformation, and if they were to be taken seriously as the template for global governance, that would quite fundamentally change how financial institutions operate. To do so would require substantial changes in the international governance of financial institutions and in the national regulation of banks, but if the SDGs are to be taken seriously as a framework for global governance, such changes need to be made in many countries.

Hence, what could a sustainability case-based strategy to address the SDGs look like? The following section will describe such a strategy. The third section discusses some existing initiatives, and the penultimate section discusses the way forward and how to ally finance more strongly with the SDGs.

SDGs strategy for the banking sector

This book chapter will address the private banking industry and not focus on public banks, such as the World Bank, the International Monetary Fund (IMF), or different domestic and multilateral development banks. These public institutions play a major role in addressing the SDGs and can be drivers or barriers for the SDGs. Many of these public financial institutions play ambivalent roles with regard to sustainable development. On the one hand, they finance and subsidize fossil fuel projects, although the World Bank was recently reconsidering its decision as to whether to invest in the last remaining coal plant in its portfolio (Reuters 2018). On the other hand, they provide guidelines for green financing, finance sustainability-related projects, and contribute to climate change funds. The discussion of the connection between these public financial institutions and the SDGs, however, would fill another book chapter on its own (on this see Horton, Chapter 13).

Beginning with an analysis of sustainability needs, banks can explore how they engage with sustainable development issues by channelling financial capital and creating innovative products, services, and strategies. The SDGs are an ideal way to identify the most important sustainability needs as banks can focus on the 17 Goals. A strategy based on the sustainability case for business can develop products and services to address the SDGs. In this section, I first define the actions that constitute banking for sustainability (i.e. the types of products and services that banks can provide), then discuss how sustainability can be assessed, and finally how it can be reported.

Banks can provide a variety products and services which are useful to address the sustainability goals. Impact investment, social banking, and SRI are products and services that provide first steps into the right direction. While SRI mainly uses sustainability indicators in order to improve financial returns, social banking and impact investment were initiated in response to sustainability needs and due to a willingness to address these needs using financial products and services. In addition to the above-named types of finance, sustainable finance might be another category that addresses environmental and societal issues.

SRI is a form of investment that uses non-financial criteria to screen out investments for social, environmental or governance reasons or to pick investments that perform well with regard to both financial and non-financial indicators (Geobey and Weber 2013). SRI – also called Responsible Investing (RI), started as a financial niche product, but found entrance into mainstream finance as well because it increases risk management in investment decisions (Weber 2015). It entails "social" screening, community investment, and shareholder advocacy (O'Rourke 2003) to guarantee sustainable financial returns. In the beginning, SRI has been used in a similar way to the current divestment (Hunt,

Weber, and Dordi 2017; Hunt and Weber 2019) approach. It mainly excluded investments, such as tobacco companies, because of ethical reasons. Newer approaches, however, used both exclusion criteria and positive selection criteria. The main goal of SRI is to achieve attractive financial returns through investments that take long-term sustainability concerns into account (Weber, Mansfeld, and Schirrmann 2011). Furthermore, SRI strives to channel financial capital towards sustainable businesses (Weber 2006; Buttle 2007). The Canadian Responsible Investment Association (RIA) provides further information on these topics on its website.[2]

In contrast to SRI, social banking and impact investing search for a positive social and financial impact. Social banks focus on achieving positive social, environmental, and sustainability impacts: they base all their business and their operations on the triple-bottom-line concept, and they use financial products and services to achieve a blended value consisting of social, environmental, and financial returns (Weber 2016b). Institutions practicing social banking are often members of the Global Alliance for Banking on Values (GABV). This association has the following principles:

1   Triple-bottom-line approach at the heart of the business model
2   Grounded in communities, serving the real economy and enabling new business models to meet the needs of both
3   Long-term relationships with clients and a direct understanding of their economic activities and the risks involved
4   Long-term, self-sustaining, and resilient to outside disruptions
5   Transparent and inclusive governance
6   All of these principles embedded in the culture of the banks (GABV 2018).

Impact investing intends to address social or environmental challenges while generating financial returns for investors. Its main goal is to create a positive societal impact through capital investment. The spectrum of financial returns can vary. Some impact investments achieve financial returns that are comparable to conventional investments, while others may not achieve financial returns. Impact investing has its origins in philanthropy. It emerged because of philanthropists trying to find ways to invest their endowments to support social or environmental issues. Impact investors typically invest in equity of social enterprises or in quasi-equity of charitable organizations or non-profits (Weber and Feltmate 2016). A main player in developing impact assessment principles, guidelines, and criteria is the Global Impact Investing Network. The network developed a comprehensive guideline and indicators to assess the impact of impact investing in a transparent way[3] (see also Chapter 13 in this book).

To assess the social, environmental, and sustainability impact of financial products, the social return on investment (SROI) can be calculated. SROI uses a set of indicators to measure the impact of a business, such as a social venture, or of a financial product, such as a social impact bond (Weber 2013a). It measures both positive and negative impacts on society (Gibson *et al.* 2011). SROI can be used to measure the impact of both products that address the SDGs and conventional products. The assessment helps to identify differences between these products and might even help to improve the social performance of a conventional product.

Banks can use impact analyses for selected products and services. If the impact analysis suggest that a project does not create sufficient sustainability returns, banks might modify or not offer the respective product. The same procedure can be applied to products and services that create negative impacts with regard to the SDGs. Furthermore, products and services with positive SROI must be in the core portfolio of sustainable banks. They cannot be niche products, because otherwise the impacts would be neutralized or even eaten up by conventional core products with neutral or negative effect for the SDGs.

Thus, the SDGs impact approach can be used for all banking products and services such as lending, asset management, and investment banking. Banks' own internal operations such as human resources can be assessed too, for instance, with regard to SDG 5 "Gender Equality" or SDG 8 "Decent Work and Economic Growth". The strong influence of banks on the business sector and economy and their high market penetration supports this strategy because the impact on the SDGs can be leveraged to borrowers, investees, and financed projects.

In addition to SROI assessment, other concepts, such as sustainability cost-benefit analyses (Rennings and Wiggering 1997), the impact reporting indicator set (IRIS) (Global Impact Investing Network 2012), sustainability balanced scorecards (Figge *et al.* 2002), environmental and carbon footprints (Wackernagel *et al.* 1999; Weidema *et al.* 2008), and key sustainability performance indicators (Epstein and Roy 2001) can be used to evaluate the sustainability impact of financial products and services. These concepts are far from perfect, but banks could play a significant role in developing them further. The application of the sustainability case for banking in turn calls for the development of appropriate indicators to assess its success.

If banks have been able to change the characteristics of the housing market by introducing asset-backed securities, and have been able to facilitate the widespread distribution of related financial products, why should they not be able to distribute sustainable finance products and services that address the SDGs? At least in developed countries, market penetration of banks and credit unions is nearly 100 per cent, giving them a great opportunity to deliver their sustainable products and services widely. In addition to retail clients, banks also cooperate with institutional investors, such as pension funds, which have strong market power. However, though many banks and credit unions offer socially responsible and impact investments, and some of them practice social banking, the market penetration of these products and services is still relatively small. The Canadian RIA reports that responsible investing, which they use as a synonym for socially responsible investing, represents 38 per cent of all managed assets. Globally, this number is 26 per cent (Global Sustainable Investment Alliance 2017). Given these numbers, we have to keep in mind that the term responsible investment is used in a broad sense, including fairly modest criteria such as Environmental, Social, and Governance (ESG) screening and shareholder engagement. This is in contrast with impact investing that addresses sustainability problems through directing investments. Impact investment in Canada represents below 1 per cent of all assets under management (Responsible Investment Association 2016).

Because many institutional investors have a long-term focus, they should be interested in a sustainability-based approach that addresses the SDGs. The Norwegian sovereign wealth fund, for instance, uses ESG criteria and started to divest from fossil fuels, addressing SDG 13 "Climate Action" (Mooney 2017). From a risk perspective, banks should be interested in supporting SDG 13. In 2015, the Governor of the Bank of England, Mark Carney, presented a speech on the "tragedy of the horizon", addressing financial risks for banks and insurers caused by climate change (Carney 2015). This speech addressed direct, regulative, reputation, and systemic risks of climate change that might have an effect on financial sector stability (Weber and Kholodova 2017). Consequently, the Task Force on Climate-related Financial Disclosures (TFCFD) (2016), a working group of the Group of Twenty (G20), whose focus is climate change and financial risks, produced a report on how the financial industry could address climate-change-related risks through establishing standardized indicators that measure the risk to their borrowers and investees. Financing SDG 13 would be a step in mitigating climate related financial risks for the banking sector (see Schweizer, Chapter 9).

Another opportunity for banks to promote the sustainability case is the connection to industries which are commercial borrowers. Banks have the ability to channel funds into sustainable industries, while also rewarding sustainability leaders and penalizing laggards through variable interest rates. These conditions make the banking industry one of the best-placed sectors to create a positive impact on sustainable development, if shareholders, managers, and directors of banks realize this opportunity to mitigate long-term financial risks. This strategy needs to permeate the whole sector: holding only specialized green and social banks as being responsible for such lending will not create enough investments to address the SDGs.

Another financing mechanism suitable to address the SDGs is project finance. Project finance is a means of finance for big projects, such as infrastructure and energy. This type of finance addresses SDG 9 directly and other SDGs, such as SDG 6, SDG 7, and SDG 11 indirectly. The current code of conduct for addressing environmental and social risks, developed by the Equator Principles Association has been adopted by about 80 project financiers, the majority of which are banks, representing the majority of global project finance (Weber and Acheta 2014; Weber 2016a). Currently, however, the Equator Principles mainly address risks. A way to address the SDGs could be to include an SDGs impact assessment into the Equator Principles that would help to identify how projects address the SDGs. Since the Equator Principles are an industry code that continuously adopts new environmental or societal developments, members of the

Equator Principles Association could update the new version of the code of conduct to address the SDGs directly.

The last link in the chain of strategizing the sustainability case for banking is reporting. With a few exceptions, sustainability reporting to date has been decoupled from financial reporting and is based on major guidelines, such as the Global Reporting Initiative (GRI) (Global Reporting Initiative 2013) including a banking sector supplement (Global Reporting Initiative 2011). Standards, such as the GRI, have already started to integrate the SDGs into their reporting guidelines. Therefore, one might expect that also banking sector organizations will integrate the SDGs into their sustainability reporting. If the SDGs reporting is conducted in an integrated way, it will connect impacts on the SDGs with financial figures and will be able to report about the benefits and risks of addressing the SDGs through financial products and services (Eccles, Cheng, and Saltzmann 2010).

If conducted in a credible way, an SDGs-based business strategy can help banks and other financial institutions regain client trust and reputation, and to re-build a positive public perception about the banking sector. Applying the sustainability case for banking presents a way to convert a weakness of the banking industry into a strength. Consequently, the proposed strategy undoes the tendency of banks to merely react to sustainability issues and replaces it with an integrated sustainability strategy. Overall, I predict this shift will create two benefits: financial support for sustainable development, and a banking sector that is transparent and that has a positive impact on society, the environment, and the economy.

Do's and don'ts with regard to the SDGs

There are a number of activities that banks already undertake to address sustainable development and the SDGs (Weber and Feltmate 2016). This section will outline some of the activities. With regard to employees, many banks and other banking sector institutions have gender policies focusing on gender equity with regard to both equal pay and diversity (SDG 5). Internally, many financial institutions are leading in energy, water, and materials management to reduce environmental impacts (SDGs 6, 11, 12, and 15). Additionally, a significant part of corporate philanthropy addresses issues such as education (SDG 4), health and well-being (SDG 3) and sustainable communities (SDG 11) (Weber and Feltmate 2016).

With regard to their products and services, the banking sector engages in green and climate bonds (SDG 7, 9, 11, and 13) (Reichelt 2010; Sustainable Prosperity 2012; Oliver and Boulle 2013; Field 2015; Kidney, Sonerud, and Oliver 2015; Weber and Saravade 2018) and invests in clean technologies (SDG 7 and 13) (Aizawa and Chaofei 2010; Arabella Advisors 2015; Cumming, Henriques, and Sadorsky 2016). Furthermore, there are social impact bonds that address social goals, such as health, education, and decent work (Mulgan *et al.* 2011; Trotta *et al.* 2015). Also on the risk management side, the financial industry addresses the SDGs by assessing the environmental and social risks of their clients and projects (Weber, Fenchel, and Scholz 2008; Weber, Scholz, and Michalik 2010; Weber, Hoque, and Islam 2015), or by divesting from activities that have negative impacts on the SDGs, such as fossil fuels on SDG 13 (Arabella Advisors 2015).

As mentioned above, these are only some examples of the financial industry addressing the SDGs. Furthermore, banks could be more active in addressing the SDGs outside of particular niche products and services. Also, in order to address the SDGs effectively, I argue that the financial industry has to reduce financing to clients and projects that have a negative impact on SDGs because the availability of financial capital is a main enabler for these clients to continue their activities (Hunt, Weber, and Dordi 2017). The financial industry might divest from industries that harm the achievement of the SDGs, such as the fossil fuel industry, industries that promote irresponsible consumption, such as the consumption of plastic that pollutes our water, or industries that harm good health and well-being, such as the tobacco industry. Another way to influence these industries is through shareholder and lender engagement to support a transition of these industries into a more sustainable direction. These approaches could be subsumed under SDG 17 "Partnerships for the Goals". Because the banking industry cannot take this step alone, it needs to collaborate with governments, central banks, and other stakeholders. China, for instance, introduced the Green Credit Guidelines to support banks in financing green industries (Cui *et al.* 2018), while others might introduce taxes, such as a carbon tax to integrate sustainability risks into financial decision-making. While there is probably not a solution that fits all, financial incentives might be more effective than trying to change mindsets (Weber 2005).

A move in this direction would enable banks to take opportunities and to avoid risks. Experience demonstrates that using a proactive strategy in addressing environmental and social issues has a positive effect on the financial bottom line of the banking industry. Lenders, for instance, integrating environmental and social risks into their commercial credit risk assessment have been able to reduce their credit risks (Weber, Scholz, and Michalik 2010; Weber, Hoque, and Islam 2015; Cui *et al.* 2018). In addition, banks that base their business strategy on addressing sustainable development have been achieving similar financial results to those of conventional banks (Weber and Remer 2011; Weber 2013b; Weber 2016b). Hence, we can expect that banks addressing sustainable development will be more successful in the long run. In addition, examples from other industries support the hypothesis that sustainability and financial returns rather go hand-in-hand than being a trade-off (Porter and Kramer 2006, 2011).

Finally, many accuse the banking sector of contributing to increasing financial inequalities (SDG 10) through the financialization of the economy (Dore 2008). Decreasing financial regulations for the big players in the field weaken the consideration that governments will be able to address these inequalities. In combination with risky behaviour in the banking industry, this might contradict SDG 10 and increase inequalities (Agnello and Sousa 2012). Therefore, one way to address SDG 10 is to de-risk the banking system through both strategy changes and regulations.

How to better connect the banking sector with the SDGs

This section will discuss ways how the banking sector could connect better with the SDGs and be incentivized to invest in sustainable development. The proposals, developed by the author of this chapter based on global approaches, address public–private partnership approaches, regulative approaches, and risk management approaches conducted by the banking industry.

First, the banking industry might develop systematic risk and opportunity assessment tools for the SDGs that they integrate into their conventional lending and investment risk management tools. A key component of this is to develop and implement an action plan for major risks and opportunities involved, including effective stakeholder communication mechanisms and risk mitigation measures. Usually banks develop these instruments in-house or they are developed by specialized rating agencies. External audits of the tools and their results might increase their transparency and credibility (Kolk 2008).

Second, aligning products and services with sustainability principles can be a business opportunity. Banks working with favourable policies with regard to the SDGs can increase their green and social finance and add SDGs-related financial products and services to their portfolios, and consequently diversify their business. Hence, also the UNEP Inquiry into the Design of a Sustainable Financial System (Zadek and Robins 2018) states that incorporating sustainability considerations, including risks and opportunities, must become a part of the financial sector's culture, business, and regulation.

Third, establishing collaboration between various global and regional public and private sector actors will be critical in addressing the SDGs, as SDG 17 emphasizes. These partnerships allow for value sharing and dissemination of expertise between stakeholders. This would help to create investments in early stage market development in lower income countries, such as investments for products and services that support sustainable development (Stafford-Smith *et al.* 2017). Public partners might also be the World Bank, and other national, regional, or multinational development banks.

Fourth, an important aspect of reporting is to embed SDGs-related indicators into the existing reporting systems. By doing so, it would make it possible to track and measure the sustainability performance of any financial institution based on the SDGs. The Global Reporting Initiative developed guidelines for integrating the SDGs in corporate reporting (Global Reporting Initiative 2018) They propose a three-step approach starting with the definition of priority SDGs targets. Banks, for instance, might define that they want to address SDG 13 through Climate Bonds and SDG 12 "Responsible Consumption and Production" through integrating ESG criteria into their lending decisions. The second step focuses setting objectives on measuring and analysing the achievements of the set objectives with regard to the prioritized SDG using data analysis. Objectives could be set using a theory of change model that defines the inputs used for a Green Bond or an ESG finance product and its corresponding impacts (Jackson 2013). The third step is to report about the strategy and results relative to the SDGs, and to present an outlook about how a bank will implement SDG-related activities into the core strategy.

Fifth, public guarantees could help the banking sector to mitigate risks with regard to SDGs finance. Instruments might be tax-related incentives and first-loss guarantees provided through a public first-loss fund. These incentives would need lower amounts of public funding than direct subsidies or public investment. Because they combine public and private finance. The World Bank's Epidemic Preparedness Bonds that are mentioned in Chapter 13 (Horton) in this book are one example. Another example is the first loss guarantees that government entities, such as the US Overseas Private Investment Corporation (OPIC), provide to incentivize foreign investment in developing countries (Goodman 2006).

Conclusion

Using a sustainability case for finance approach could help the banking sector to address the SDGs in an effective way. This approach would be a win-win for both the SDGs and the financial industry that will be able to mitigate risks and to take business opportunities. The existing codes of conducts and reporting guidelines in the industry are already a good basis for addressing the SDGs. However, they should be expanded using a positive approach instead of focusing on the management of social, environmental, and sustainability risks. A first step into this direction comes from the UN Environment Programme Financial Initiative (UNEP FI) that developed its positive impact initiative that explicitly addresses the SDGs.[4]

But clearly much more needs to be done to convert the proposals outlined in this chapter into practical measures by the global community to implement sustainability across the financial sector, to, in Zadek and Robins' (2018) terms, "make waves" to change how finance works in promoting sustainability. The SDGs imply that global governance has to be more comprehensively implemented in social arenas normally considered outside the operation of international relations. While forcing banks and other institutions to reconsider how they make their investment decisions may appear draconian, given the rising financial risks from climate change in particular and failure to think about sustainability more generally, such regulatory guidelines ought to improve the performance of the financial sector in monetary terms. This is, as this chapter outlines, not inconsistent with many nascent initiatives in the sector, and hence there are inherent incentives in favour of such regulations which ought to make this task possible in many venues. If this is done, then the chances of successful implementation of many of the SDGs will clearly be improved.

Notes

1 For information on RBC's Diversity and Inclusion police, please see www.rbc.com/history/
   milestones-at-a-glance/diversity-and-inclusion.html. Accessed 21 November 2018.
2 Please see the following link for more on RIA: www.riacanada.ca/about/. Accessed 21
   November 2018.
3 See www.iris.thegiin.org. Accessed 21 November 2018.
4 More information on UNEP FI's Positive Impact Initiate can be found at: www.unepfi.
   org/positive-impact/positive-impact/. Accessed 21 November 2018.

References

Agnello, L., and R. M. Sousa. 2012. "How do Banking Crises Impact on Income
   Inequality?". *Applied Economics Letters* 19 (15): 1425–1429.
Aizawa, M., and Y. Chaofei. 2010. "Green Credit, Green Stimulus, Green Revolution?
   China's Mobilization of Banks for Environmental Cleanup". *The Journal of Environ-
   ment and Development* 19 (2): 119–144.
Arabella Advisors. 2015. *Measuring the Growth of the Global Fossil Fuel Divestment and
   Clean Energy Investment Movement*. Washington, DC: Arabella Advisors.
Buttle, M. 2007. "'I'm Not in It for the Money': Constructing and Mediating Ethical
   Reconnections in UK Social Banking". *Geoforum* 38 (6): 1076–1088.
Carney, M. 2015. "Breaking the Tragedy of the Horizon: Climate Change and Financial
   Stability". Speech given at Lloyd's of London by the Governor of the Bank of England,
   29 September.
Cui, Y., S. Geobey, O. Weber, and H. Lin. 2018. "The Impact of Green Lending on
   Credit Risk in China". *Sustainability* 10, 2008: 1–16.
Cumming, D., I. Henriques, and P. Sadorsky. 2016. "'Cleantech' Venture Capital around
   the World". *International Review of Financial Analysis* 44: 86–97.
Dore, R. 2008. "Financialization of the Global Economy". *Industrial and Corporate Change*
   17 (6): 1097–1112.
Dyllick, T., and K. Hockerts. 2002. "Beyond the Business Case for Corporate Sustain-
   ability". *Business Strategy and the Environment* 11: 130–141.
Eccles, R. G., B. Cheng, and D. Saltzmann, eds. 2010. *The Landscape of Integrated Report-
   ing Reflections and Next Steps*. Cambridge, MA: The President and Fellows of Harvard
   College.
Epstein, M. J., and M.-J. Roy. 2001. "Sustainability in Action: Identifying and Measuring
   the Key Performance Drivers". *Long Range Planning* 34 (5): 585–604.
Field, A. 2015. "$36.6B In Green Bonds Issued Last Year". *Forbes*, 15 January. New York:
   Forbes.
Figge, F., T. Hahn, S. Schaltegger, and M. Wagner. 2002. "The Sustainability Balanced
   Scorecard: Linking Sustainability Management to Business Strategy". *Business Strategy
   and the Environment* 11 (5): 269–284.
GABV (Global Alliance for Banking on Values). 2018. "Principles". Accessed 20
   November 2018. www.gabv.org/about-us/our-principles.
Geobey, S., and O. Weber. 2013. "Lessons in Operationalizing Social Finance: The Case
   of Vancouver City Savings Credit Union". *Journal of Sustainable Finance and Invest-
   ment* 3 (2): 124–137.
Gibson, J., A. Jones, H. Travers, and E. Hunter. 2011. "Performative Evaluation and
   Social Return on Investment: Potential in Innovative Health Promotion Interven-
   tions". *Australasian Psychiatry* 19 (1): S53–S57.

Global Impact Investing Network. 2012. "IRIS Metrics". New York, NY: Global Impact Investing Network. Accessed 5 December 2012. http://iris.thegiin.org/about-iris.

Global Reporting Initiative. 2011. *Sustainability Reporting Guidelines and Financial Services Sector Supplement*. Amsterdam, The Netherlands: The Global Reporting Initiative.

Global Reporting Initiative. 2013. *G4 Sustainability Reporting Guidelines*. Amsterdam, The Netherlands: Global Reporting Initiative.

Global Reporting Initiative. 2018. *Integrating the SDGs into Corporate Reporting: A Practical Guide*. Amsterdam, The Netherlands: Global Reporting Initiative and UN Global Compact.

Global Sustainable Investment Alliance. 2017. *2016 Global Sustainable Investment Review*. NA, Global Sustainable Investment Alliance.

Goodman, P. 2006. "Microfinance Investment Funds: Objectives, Players, Potential". In *Microfinance Investment Funds: Leveraging Private Capital for Economic Growth and Poverty Reduction*, edited by I. Matthäus-Maier, and J. D. Von Pischke, 11–45. Berlin: Springer.

Hunt, C., and O. Weber. 2019. "Fossil Fuel Divestment Strategies: Financial and Carbon-Related Consequences". *Organization and Environment* 32 (1): 41–61.

Hunt, C., O. Weber, and T. Dordi. 2017. "A Comparative Analysis of the Anti-Apartheid and Fossil Fuel Divestment Campaigns". *Journal of Sustainable Finance and Investment* 7 (1): 64–81.

Jackson, E. T. 2013. "Interrogating the Theory of Change: Evaluating Impact Investing Where It Matters Most". *Journal of Sustainable Finance and Investment* 3 (2): 95–110.

Kidney, S., B. Sonerud, and P. Oliver. 2015. *Growing a Green Bonds Market in China: Key Recommendations for Policymakers in the Context of China's Changing Financial Landscape*. Ottawa, ON: International Institute for Sustainable Development.

Kolk, A. 2008. "Sustainability, Accountability and Corporate Governance: Exploring Multinationals' Reporting Practices". *Business Strategy and the Environment* 17 (1): 1–15.

Monteiro, C. A., J. C. Moubarac, G. Cannon, S. W. Ng, and B. Popkin. 2013. "Ultra-processed Products are Becoming Dominant in the Global Food System". *Obesity Reviews* 14 (2): 21–28.

Moodie, R., D. Stuckler, C. Monteiro, N. Sheron, B. Neal, T. Thamarangsi, P. Lincoln, S. Casswell, and L. N. A. Group. 2013. "Profits and Pandemics: Prevention of Harmful Effects of Tobacco, Alcohol, and Ultra-processed Food and Drink Industries". *The Lancet* 381 (9867): 670–679.

Mooney, A. 2017. "Growing Number of Pension Funds Divest from Fossil Fuels". *Financial Times*, 28 April. Accessed 21 November. www.ft.com/content/fe88b788-29ad-11e7-9ec8-168383da43b7.

Mulgan, G., N. Reeder, M. Aylott, and L. Bo'sher. 2011. *Social Impact Investment: The Challenge and Opportunity of Social Impact Bonds*. London: The Young Foundation.

Oliver, P., and B. Boulle. 2013. *Bonds and Climate Change: The State of the Market in 2013*. London: Climate Bonds Initiative.

O'Rourke, A. 2003. "The Message and Methods of Ethical Investment". *Journal of Cleaner Production* 11 (6): 683–693.

Porter, M. E., and M. R. Kramer. 2006. "Strategy and Society: The Link Between Competitive Advantage and Corporate Social Responsibility". *Harvard Business Review* 84: 78–92.

Porter, M. E., and M. R. Kramer. 2011. "Creating Shared Value". *Harvard Business Review* 89: 62–77.

Reichelt, H. 2010. "Green Bonds: A Model to Mobilise Private Capital to Fund Climate Change Mitigation and Adaptation Projects". *The EuroMoney Environmental Finance Handbook*: 1–7.

Rennings, K., and H. Wiggering. 1997. "Steps Towards Indicators of Sustainable Development: Linking Economic and Ecological Concepts". *Ecological Economics* 20 (1): 25–36.

Responsible Investment Association. 2016. *2016 Canadian Impact Investment Trends Report*. Toronto: Responsible Investment Association.

Reuters. 2018. "World Bank in Doubt Whether to Back Kosovo Coal-fired Power Plant". *Reuters*, 13 June. Accessed 16 September 2018. https://uk.reuters.com/article/kosovo-worldbank-energy/world-bank-in-doubt-whether-to-back-kosovo-coal-fired-power-plant-idUKL8N1TF4GW.

Schaltegger, S., and R. Burritt. 2015. "Business Cases and Corporate Engagement with Sustainability: Differentiating Ethical Motivations". *Journal of Business Ethics* 147 (2): 241–259.

Stafford-Smith, M., D. Griggs, O. Gaffney, F. Ullah, B. Reyers, N. Kanie, B. Stigson, P. Shrivastava, M. Leach, and D. O'Connell. 2017. "Integration: The Key to Implementing the Sustainable Development Goals". *Sustainability Science* 12 (6): 911–919.

Sustainable Prosperity. 2012. "Green Bonds". In *Sustainable Prosperity*. Policy Brief. June. Ottawa: Sustainable Prosperity.

Trotta, A., R. Caré, R. Severino, M. C. Migliazza, and A. Rizzello. 2015. "Mobilizing Private Finance for Public Good: Challenges and Opportunities of Social Impact Bonds". *European Scientific Journal* 1: 259–279.

Wackernagel, M., L. Onisto, P. Bello, A. C. Linares, I. S. L. Falfán, J. M. García, A. I. S. Guerrero, and G. S. Guerrero. 1999. "National Natural Capital Accounting with the Ecological Footprint Concept". *Ecological Economics* 29 (3): 375–390.

Weber, O. 2005. "Sustainability Benchmarking of European Banks and Financial Service Organizations". *Corporate Social Responsibility and Environmental Management* 12 (2): 73–87.

Weber, O. 2006. "Investment and Environmental Management: The Interaction between Environmentally Responsible Investment and Environmental Management Practices". *International Journal of Sustainable Development* 9 (4): 336–354.

Weber, O. 2013a. "Impact Measurement in Microfinance: Is the Measurement of the Social Return on Investment an Innovation in Microfinance?" *Journal of Innovation Economics and Management* 11 (1): 149–171.

Weber, O. 2013b. "Social Banks and Their Profitability: Is Social Banking in Line with Business Success?" In *Prospective Innovation at Ethical Banking and Finance*, edited by L. San-Jose, and J. L. Retolaza, 2–19. Sumy: Vinnychenko M.D.

Weber, O. 2014. "The Financial Sector's Impact on Sustainable Development". *Journal of Sustainable Finance and Investment* 4 (1): 1–8.

Weber, O. 2015. "Sustainable Finance". In *Sustainability Science Handbook*, edited by H. Heinrichs, P. Martens, G. Michelsen, and A. Wiek, 119–127. Heidelberg: Springer.

Weber, O. 2016a. "Equator Principles Reporting: Factors Influencing the Quality of Reports". *International Journal of Corporate Strategy and Social Responsibility* 1: 141–160.

Weber, O. 2016b. "Social Banks' Mission and Finance". In *Routledge Handbook on Social and Sustainable Finance*, edited by O. M. Lehner, 467–479. London: Routledge.

Weber, O., and S. Remer, eds. 2011. *Social Banks and the Future of Sustainable Finance*. London: Routledge.

Weber, O., and E. Acheta. 2014. "The Equator Principles: Ten Teenage Years of Implementation and a Search for Outcome". *CIGI Papers Series*. Waterloo, ON: CIGI.

Weber, O., and B. Feltmate. 2016. *Sustainable Banking and Finance: Managing the Social and Environmental Impact of Financial Institutions.* Toronto: University of Toronto Press.

Weber, O., and O. Kholodova. 2017. *Climate Change and the Canadian Financial Sector.* Waterloo, ON: Centre for International Governance Innovation (CIGI).

Weber, O., and V. Saravade. 2018. "Green Bonds are Taking Off – and Could Help Save the Planet". *The Conversation and National Post*, 4 January.

Weber, O., M. Fenchel, and R. W. Scholz. 2008. "Empirical Analysis of the Integration of Environmental Risks into the Credit Risk Management Process of European Banks". *Business Strategy and the Environment* 17 (3): 149–159.

Weber, O., R. W. Scholz, and G. Michalik. 2010. "Incorporating Sustainability Criteria into Credit Risk Management". *Business Strategy and the Environment* 19 (1): 39–50.

Weber, O., M. Mansfeld, and E. Schirrmann. 2011. "The Financial Performance of RI Funds After 2000". In *Responsible Investment in Times of Turmoil*, edited by W. Vandekerckhove, J. Leys, K. Alm, B. Scholtens, S. Signori, and H. Schaefer, 75–91. Berlin: Springer.

Weber, O., A. Hoque, and A. M. Islam. 2015. "Incorporating Environmental Criteria into Credit Risk Management in Bangladeshi Banks". *Journal of Sustainable Finance and Investment* 5 (1–2): 1–15.

Weidema, B. P., M. Thrane, P. Christensen, J. Schmidt, and S. Løkke. 2008. "Carbon Footprint". *Journal of Industrial Ecology* 12 (1): 3–6.

Wiek, A., and O. Weber. 2014. "Sustainability Challenges and the Ambivalent Role of the Financial Sector". *Journal of Sustainable Finance and Investment* 4 (1): 9–20.

Zadek, S., and N. Robins. 2018. *Making Waves – Aligning the Financial System with Sustainable Development.* Geneva: United Nations Environment Programme.

15 Global governance and the Sustainable Development Goals

Alexandra R. Harrington

Introduction

When the Balsillie School of International Affairs was founded in 2008, the Sustainable Development Goals (SDGs) were unknown in the international community, which was focused instead on achieving the Millennium Development Goals (MDGs) by 2015. Over the course of the past ten years, as there has been a significant shift in the focus from the vast yet difficult to quantify MDGs to the more tailored and quantifiable SDGs, there has been a concomitant shift in the ways in which global governance has been incorporated by – and incorporates – how the international community seeks to address the most fundamental challenges before it. These shifts demonstrate the ways in which global governance structures are essential to the achievement of the SDGs and how the SDGs, in turn, provide unique opportunities for global governance mechanisms in the present moment and in the future.

This chapter examines the duality existing between the SDGs and global governance, notably the global governance mechanisms that are essential for implementing the work of the international community and, often, the regional community. The second section, sets out the provisions of the SDGs, and the targets through which they are to be measured and achieved, that have incorporated aspects of global governance generally and global governance mechanisms in particular. Mapping these connections reveals many fascinating areas of intersection and overlap that exist overtly. Beyond this, such a mapping exercise demonstrates the importance of seeing where there are tacit intersections between the SDGs and global governance. Without understanding the roles which these connections play overtly and tacitly it is otherwise difficult to understand the full width and breadth of the SDGs and, perhaps more fundamentally, of the concept and application of global governance in a flexible manner which validates its continued importance for current and future generations.

The third section of this chapter then discusses the ways in which the SDGs impact on and can drastically change the functions and aims of global governance mechanisms. To do so, the section sets out three genres of global governance mechanism or evaluation. These genres – newly created mechanisms for the SDGs themselves, financial mechanisms, and environmental mechanisms – are essential to grasping the scope of the SDGs' impacts. Within each section, examples of prominent global governance mechanisms and their intersections with the SDGs are offered in order to provide an understanding of how these mechanisms have changed and adapted in order to incorporate and achieve the SDGs. These understandings also demonstrate the ways in which various constituencies served by these global governance mechanisms have had their interests changed as a result of the SDGs.

Finally, the fourth section examines the implications of the relationship between global governance and the SDGs, highlighting lessons learned and potential avenues of growth and change in the future. It concludes that this relationship will continue to grow and change the nature of both entities at a robust and fundamental level for years to come. Although the SDGs are set to expire in 2030, the impacts they have had and continue to have on global governance mechanisms will undeniably exist well beyond this point.

SDGs and avenues for global governance

At the turn of the millennium, the international community came together to acknowledge the eight most pressing issues facing global society and to create broad-based goals for addressing them (UNGA 2000). Rather than fashioning open-ended goals in terms of temporal requirements, these MDGs instead were given the finite duration from 2000 to 2015 (UNGA 2000).

Throughout these 15 years, a number of steps were taken and certain aspects of the MDGs were achieved, although many remained incomplete (UNGA 2015). Regardless the success rate and overall tangible results, the MDGs combined to change the ways in which the international community identifies the most pressing issues of our time and creates and implements methods for them to be addressed. Through the vehicle of the MDGs, the many silos in which various aspects of international law, policy, and development operate began to be weakened, replaced instead with a more expanded understanding of the ways in which these are interrelated.

Work on the post-MDGs system began well in advance of 2015, and featured attempts to identify the ways in which the MDGs were successful – or not – in terms of their intended functions (UNGA 2015). At the same time, the international community sought to ensure that the next step onward from the MDGs addressed the most pressing issues of the time rather than within the uncertainty of a changing millennium (UNGA 2015).

Set against this backdrop, the United Nations entity ultimately tasked with devising the MDGs' successors identified sustainable development as the prism through which to address the most pressing issues to the international community and which were likely to continue to be pressing – and in some cases, such as climate change, only increase in importance – in the future (UNGA 2015). At the same time, the crafters of the post-MDG regime determined that it was necessary to identify more precise goals for the next time period along with targets to further elucidate their scope and indicators to serve as methodological tools for determining the success of the goals and targets (UNGA 2015).

What ultimately emerged were the 17 SDGs, accompanied by over 150 targets and even more indicators, which are to remain in effect and are to be accomplished by 2030 (UNGA 2015). The SDGs touch on a myriad of topics and fields, ranging from poverty, to health, to climate to peace. One of the threads which runs through them and through concepts of the ways in which they are to be implemented is the role of global governance.

In SDG 1 on eliminating poverty, there are a number of national requirements, benchmarks and indicators (UNGA 2015). This does not, however, exclude the vital role of global governance mechanisms, as under Target 1.b there is the requirement that at all levels, including the global level, pro-poor and anti-poverty frameworks be created, implemented, and overseen with the goal of ending poverty by 2030. Global governance mechanisms and systems have an additional role to play under Target 1.a, in which there is a requirement that resources be mobilized across a spectrum of scales and needs to effect the implementation of poverty elimination plans.

Concomitant with this, SDG 2 on hunger has broad-based national and international targets and obligations that require a level of global governance and oversight that entails issues ranging from food security to nutrition and beyond (UNGA 2015). For the protection and maintenance of plants and biodiversity, for example, Target 2.5 provides for the establishment of international agreements and standards on conservation and protection that require national and international governance oversight in order to properly reflect the gravity of these issues at multiple levels. In connection with food and hunger, access to water, addressed in SDG 6, includes significant international co-ordination efforts and the understanding that international action will be necessary in areas such as capacity building to encourage the achievement of the targets and indicators (UNGA 2015). Water, as a natural resource and an essential commodity of life, is subject to global governance throughout the SDGs, especially SDG 6, which requires support and co-ordination of water-related scaling up and other activities, as well as a host of national activities.

Health, as addressed under SDG 3, contains many forms of health issue areas, communicable and non-communicable, all with a general focus on the implementation of protections and improvements by national governments (UNGA 2015). In addition, there is an emphasis on implementing international laws relating to or impacting health, such as the World Health Organization (WHO) Framework Convention on Tobacco Control Convention (Target 3.a) and the binding International Health Regulations (Target 3.d). Further, Target 3.b includes various World Trade Organization (WTO) binding laws, such as TRIPS Agreement (on Trade-Related Aspects of Intellectual Property Rights), within the ambit of health policy goals such as vaccination access, especially for those in developing states, and the furtherance of research and development for vaccination and medication resources.

The equality and protection of women, and particularly girl children, as contained in SDG 5, requires a broader system of international implementation and oversight as a whole (UNGA 2015). This is notably reflected in the requirement that states act in compliance with the Programme of Action of the International Conference on Population and Development, along with the Beijing Platform of Action, which have established methods of oversight and global governance implementation.

Innovation for addressing climate change and advancing sustainable development features prominently throughout the SDGs and incorporates global governance in many ways. In SDG 7, relating to energy and energy modernization, there is a specific requirement that the international community co-operate "to facilitate access to clean energy research and technology, including renewable energy, energy efficiency and advanced and cleaner fossil-fuel technology, and promote investment in energy infrastructure and clean energy technology" (UNGA 2015, 19). The implementation of this aspect of SDG 7 directly relates to global governance mechanisms and their development of systems to facilitate the achievement of this goal. In terms of economic growth, SDG 8 establishes the need for an internationally focused progress in "resource efficiency in consumption and production and endeavour to decouple economic growth from environmental degradation, in accordance with the 10-Year Framework of Programmes on Sustainable Consumption and Production, with developed countries taking the lead" (UNGA 2015, 19). In itself, this puts global governance mechanisms at the forefront of the SDGs, as such requirements can only be meaningfully implemented through a robust global governance system. At the same time, SDG 8's targets require that there be an increase in funding for and application of international Aid for Trade systems, such as those used by the WTO. This inherently mandates the existence of a meaningful global governance system through which to implement, oversee, and develop such programs in order to meet the evolving needs of trade-based aid.

Infrastructural focus within the SDGs is an essential backdrop, however it is particularly essential in SDG 9. In the terms of this SDG there is an inherent, as well as overt, understanding that infrastructure must be conceived of in a transboundary and regional context for it to be developed in a meaningful and durable way. The need to share knowledge as well as resources in order to promote infrastructural development in developing states is further mentioned in SDG 9, underscoring the ways in which concerted global governance mechanism activity is essential to achieving resiliency (UNGA 2015).

Global governance mechanisms, and efforts to make them more inclusive and thus more meaningful, are at the heart of SDG 10 on inequalities. Indeed, Target 10.6 sets out clear requirements for international organizations and global governance mechanisms to "ensure enhanced representation and voice for developing countries in decision-making in global international economic and financial institutions in order to deliver more effective, credible, accountable and legitimate institutions" (UNGA 2015, 21). Similarly, there is a requirement that states "implement the principle of special and differential treatment for developing countries, in particular least developed countries, in accordance with World Trade Organization agreements" (21) which again high-lights the role of international organizations and, more specifically, global governance mechanisms, in the achievement of inequality reduction. Further, the need for global governance mechanisms to provide appropriate assistance is evident in Target 11. 5, which requires that there be a reduction in

> the number of deaths and the number of people affected and substantially decrease the direct economic losses relative to global gross domestic product caused by disasters, including water-related disasters, with a focus on pro-tecting the poor and people in vulnerable situations.
>
> (UNGA 2015, 22)

Achieving this would be inherently impossible without a robust global governance mechanism.

The SDGs further remind and emphasize to the international community the importance of world cultural and natural heritage as part of a broader under-standing of humanity and the role of culture and heritage for future generations (Target 10.4). In order to accomplish this, the SDGs emphasize the role of international standards for the preservation of cultural and natural heritage – such as those developed under the auspices of the United Nations Educational, Cultural and Scientific Organization (UNESCO) – and the governance mecha-nisms necessary to accomplish this. In connection with inequalities, consump-tion practices and policy is another area of SDGs focus in which internationally crafted aims and practices, which are subject to global governance mechanisms, are included at the fundamental level (Target 12.1).

Given the nature of climate change as an essentially international issue, it is perhaps not a surprise that SDG 13 on climate change contains a number of intersections with global governance mechanisms, such as assisting in the imple-mentation of adaptation measures (SDG 13, Target 13.1, and Target 13.3). Indeed, SDG 13 explicitly states that it is to be viewed in light of the United Nations Framework Convention on Climate Change (UNFCCC) and subject to it as "the primary international, intergovernmental forum for negotiating the global response to climate change" (UNGA 2015, 23 [footnote]). In connection with preventing the impacts of climate change, SDG 14 on life below water includes the WTO's system – inherently one of global governance – as a site of policy development and goal implementation (Target 14.6). Terrestrial aspects of the SDGs include the requirement of global co-ordination and action to reduce deforestation and increase reforestation (Target 15.2), as well as address-ing issues of desertification and drought (Target 15.3).

Within the SDGs there is perhaps no more fundamental element that demonstrates the inherent connection between the SDGs and global govern-ance than Target 16.3, which succinctly provides that states are to "promote the rule of law at the national and international levels and ensure equal access to justice for all" (UNGA 2015, 25). Establishing the rule of law at any level – international, national, or sub-national – involves the use of governance mechanisms and systems for definition, oversight, and future development. Thus, at the level of the SDGs, global governance mechanisms are vital for the creation of effective rule of law strategies without which the rest of the SDGs might not be possible to achieve. This is supported by Target 16.8, requiring the bolstering of developing State activities and participation in global governance and associated mechanisms. Additionally, SDG 17 demonstrates the necessity for global governance mechanisms and their inherent place in the implementation and achievement of the goals as a whole.

Overall, the above has highlighted the ways in which the SDGs incorporate global governance mechanisms. In this context, it is essential to note that there are overt and tacit areas of connection and overlap, and that any reading of the SDGs must include this as a basic understanding point.

Influences of SDGs on global governance

As is clear from the above, global governance has many roles to play in the fostering and implementation of the SDGs at the international and national levels. Some of these roles are quite obvious, indeed some are directly referenced, establishing certain international organizations and other entities charged with global governance mandates as benchmarks for the implementation and achievement of the SDGs and their targets.

This assumes that such organizations and entities have the ability to work at the scales necessary for and envisioned by those who saw them as dynamic and diverse enough to handle the myriad of issues posed by the SDGs. At the same time, as also noted, the SDGs intend to work with, and are influenced by, global governance mechanisms at a more subtle level. This level exists through implication and through the need for criteria, expertise, co-ordination, and oversight that only global governance mechanisms are capable of providing at the range of scales necessary to implement and achieve the SDGs.

However, the assumption that influence flows only one way in the relationship between the SDGs and global governance mechanisms is erroneous at best and at worst overlooks the ability of global governance mechanisms to function as dynamic and responsive entities themselves. What is compelling in this context is that, on examination, the SDGs offer ways for global governance mechanisms to evolve in meaningful ways to their overall missions – be they treaty-based, international organization based, or a combination of both – and guarantee their continued relevance and legitimacy. Also striking is the truth of this relationship across a broad sector of global governance mechanisms, regardless of area of expertise, scope, genre, or intent.

Nearly every international organization and global governance mechanism is in some way impacted by the SDGs, and a complete study of these relationships is beyond the scope of the present chapter. Below, however, is a profile of some of the most prominent global governance mechanisms in relation to the SDGs and how they have been directly impacted by the terms and requirements of the SDGs in their operations and, more fundamentally, in their identity.

New global governance mechanisms

The SDGs were created through the auspices of the United Nations and under the direct influence of the UN General Assembly (UNGA 2015). In many ways this was arguably one of the few bodies capable of convening the voices and actors necessary to craft a set of goals which reflects the broad scope of needs and areas of concern for the global community, as well as the inherent flexibility necessary for these goals to be implemented and made effective at the national and even sub-national levels. In order to maintain this form of connection to an international entity with a broad scope and with significant levels of legitimacy, the UNGA created a separate division within the United Nations Department of Economic and Social affairs (UNDESA) that is charged solely with handling the SDGs, their implementation, development, and oversight (United Nations 2018).

Through its mandate, the Division has assumed secretariat functions for the SDGs while at the same time taking up a global governance mantle in terms of receiving reports from states, and other entities regarding SDG implementation, providing analysis of these reports, and managing the ways in which UN agencies themselves work together on SDG implementation as part of the wider ambit of the UN as an international organization (United Nations 2018). In this way, the Division can be seen as functioning to ensure that a global governance function exists for many of the organizations handling the implementation of the SDGs as well as for the implementation of the SDGs themselves, establishing a dual-layer construct of governance that is unique within the UN system and the way in which global governance mechanisms are often conceived.

As this system is fairly new and the full scope of its powers is being defined, there is certainly the potential for growth in terms of Division oversight and policy roles. Once the first full set of global Voluntary National Reviews (VNRs) has been received, for example, it is possible that the Division could assert a greater voice in defining the acceptable parameters for how the SDGs have been implemented to date and will be implemented in the future, assuring it a significant policy role in global governance in addition to the creation and implementation of the system through which the VNRs are received and reviewed (High Level Political Forum on Sustainable Development 2018). Further, while the Division currently manages much of its UN agency coordination functions through an informal mechanism – perhaps essential given the emerging and often fluid ways in which the SDGs are understood – it is possible for this to become a far more concretized and mandatory mechanism through which the Division could create and implement governance policy and activity. Indeed, this could be extended to include UN-affiliated agencies as well should it be deemed appropriate. Regardless, the ability to establish the

Division as the global governance mechanism hub for the SDGs themselves exists for the future and demonstrates the ways in which the SDGs can give rise to new forms of global governance.

Further, when the SDGs were created the UNGA placed the United Nations Economic and Social Council (ECOSOC) in charge of establishing and overseeing the High Level Political Forum (HLPF) on Sustainable Development (UNGA 2015). The intension of the HLPF is drastically different than that of the Division in that it implements its functions over the course of a finite period of less than two weeks every year during which it convenes the HLPF at the United Nations. This function is a combination of ministerial meetings regarding the implementation and achievement of selected SDGs each year and meetings of civil society organizations, non-governmental organizations, intergovernmental organizations, and others involved in working on the designated SDGs. The political function of the HLPF extends to the ministerial conference's ability to create political declarations relating to the SDGs during the annual meeting.

The HLPF serves a significant role as functionally the equivalent to a Conference of the Parties system used for other international treaties, while at the same time being quite limited in terms of overall governance abilities. This is not to take away from the potential importance of the HLPF as a source of global governance in a softer law sense. Nor is it to minimize the unique role of the HLPF within the SDG governing structure, as the Conference of the Parties function is vital to establishing the areas of gaps and innovations within the SDGs governance structure.

Financial mechanisms

As is evident, the SDGs are truly multi-faceted and recognize the need for financial mechanisms and organizations to be heavily involved in bringing about their aims and targets. Often, the SDG targets place stress on the national governments in order to ensure that they allocate state spending and funding in ways that provide assistance to the implementation of the goals themselves.

In addition to setting national spending priorities and requiring states to directly report on how these have been and are being met throughout the life-cycle of the SDGs, there is an implicit requirement that international funding plays a key role in the implementation of the SDGs. Some instances of this place the burden of financing directly on developed states through overseas aid-based systems. These systems are potentially beneficial but come with a host of issues, such as changes in the political and financial climate of the developed state, which can profoundly change the ways that the SDGs are funded. This topic is relative to systems of national funding and governance, and thus beyond the scope of this chapter.

Beyond this, however, there remains a significant role and place for regional and international financial systems as sites of global governance mechanisms that are impacted by the SDGs. While the SDGs might be impacted by the existence of these financial mechanisms in terms of planning and allocation of burdens and asset generation, the ways in which the SDGs as a whole – even those not specifically addressing financial issues – have impacted and continue to impact financial mechanisms and their means of global governance is profound.

World Trade Organization

In the SDGs there are references to strengthening the use of Aid for Trade programs as part of the methodology used to implement the SDGs per se. One of the forerunners in the Aid for Trade program concept at the global governance level was the World Trade Organization's similarly named apparatus, which has been in place for a number of years and has become entrenched in the WTO system. Subsequently, the WTO has used the program as a benchmark in itself and as a method to assist developing states as they transition to full membership in the organization and all the financial impacts this implies (WTO 2018a). Overall, the WTO views trade as an essential aspect for achieving or promoting the accomplishment of the majority of SDGs, including anti-poverty, health, equality, work, and environmental aspects (WTO 2018b).

This not only means that the WTO will assist in the implementation of the SDGs, it also means that the WTO will – and has already – framed its work and policy platforms in ways that reflect changed understandings of how trade and the SDGs are related and what this means for the global governance systems at work in the WTO (WTO 2018b). While this has certainly been the case in the way that the WTO's periodic negotiations operate, for example through the Doha negotiation rounds since 2015, it has also become the case for the WTO's Dispute Settlement Body, which has increasing been influenced by the underlying policy goals and values found in the SDGs. This is a fundamental shift from the passivity with which the WTO defined its roles and responsibilities *vis a vis* the MDGs, where the WTO considered that the extent of its roles and obligations was less and where it essentially viewed itself as one of the global governance mechanisms charged with monitoring the relationship between the SDGs and trade (WTO 2015).

Indeed, the WTO (2017, 3) has stated that "efforts to strengthen and reform the WTO can help to support efforts at the national level and ensure that the benefits of trade are spread more widely," which directly implicates the SDGs and notes their importance as tools of global governance mechanisms in the financial sector. Further, the WTO has noted that reforms to existing systems such as TRIPS in order to reflect the requirements of the SDGs and the increase in the use and place of other governance systems intended to assist developing states as well as the trading system overall have had a significant impact on reforming the WTO as a whole and "show that the system is adaptable and dynamic in its response to the changing landscape and emerging challenges" (WTO 2017, 3). At the same time, the WTO has identified a number of areas in which it needs to change policies and practices as an organization and as a global governance mechanism in order to properly and thoroughly implement the SDGs to the extent that it is capable. These are long-term changes, impactful for current and future generations in all states – developed and developing – and will doubtless change the ways in which the WTO governs as well as the ways in which it is viewed as an organization.

World Bank

As a major source of international funding, the World Bank has a recognized and pivotal role in the implementation of the SDGs. Beyond this, however, the SDGs have begun to alter the ways in which the World Bank Group as a whole views lending and the ways in which it engages in lending operations and priorities. For example, the World Bank Group has recently announced the implementation of new bond measures through the International Development Association (IDA) that will assist in the development and implementation of funding projects that assist extremely poor states, including those that are recovering from conflict and severe issues such as natural disasters (World Bank Group 2018).

Additionally, the World Bank Group is now creating entirely new classes of bonds and financing mechanisms which are directly linked to the SDGs as well as so called "green bonds", which promote funding for climate change and similarly focused activities (World Bank Group 2018). Further, the World Bank Group has created a new entity, the Pandemic Emergency Financing Facility, which is geared towards assisting states suffering and/or recovering from severe pandemic emergencies. These are pressing issues of our time that impact on current and future generations and are vital for accomplishment of the SDGs. At the same time, they reflect a significant shift in the way that the World Bank Group envisions itself and its functions, moving from a lender to a source of financing support to combat key issues for the long and short term.

They also reflect a concomitant change in the ways that the World Bank Group sees and values risk, as the states and situations for which it is now providing financing and support represent some of the weakest and least financially stable. Further, the World Bank Group has begun to invest in, promote, and recognize the importance of data creation and mining for development at national and international levels, and, as a result, it has begun to implement methods through which to assist states and its own organizational operations to incorporate data into their planning and operational phases (World Bank Group 2018).

In addition, the World Bank Group has expanded its field of partner organizations and entities as it works to implement the aspects of the SDGs with which it is tasked. This has the effect of expanding the scope of the World Bank Group's work and knowledge base while, in cases where it is part of a group or board which evaluates issues relating to SDGs and specific financial and social aspects, making it subject to a separate and outside form of global governance mechanism, even if this is an informal one (World Bank Group 2018).

Regional development banks

Large international banks and banking systems do not hold a monopoly on global governance in the field of finance and financial mechanisms. Indeed, over the course of recent history, regional development banks have served as sources of innovation in governance techniques and systems that have become emulated across fields and mechanisms. For example, the Asian Development Bank's creation of stringent anti-corruption methodology and requirements for its lending mechanisms was at the forefront of efforts to address anti-corruption globally as well as regionally (Asian Development Bank 2017).

The Asian Development Bank has committed itself to the SDGs through the incorporation of them into its 2017–2020 Framework for Operations and Corporate Policy (Asian Development Bank 2017). In this way, it has pledged to ensure that the SDGs are part of its global governance function at the regional level, which in turn has the potential to create impacts far beyond the dimensions of its regional focus. Further, the Asian Development Bank has announced that it will undertake financing shifts similar to those used by the World Bank Group in terms of identifying areas of sensitivity which are traditionally not open to financing – such as climate change impacts or post-disaster situations – and including them in more specialized bonds and other forms of funding apparatuses (Asian Development Bank 2018).

Similarly, the Inter-American Development Bank (IADB) has recognized the importance of the SDGs to the overall functions it was created to address and to the funding schemes it implements (IADB 2017). The impact of the SDGs and their attendant emphasis on sustainability can perhaps be best observed in the IADB's overall shift in operations toward promoting and preserving sustainability as one of its core objectives and requirements for funding activities and decision making (IADB 2017).

Noting the heavy impact of climate change and associated issues in the Inter-American region, the IADB has extended the scope of its operations to include non-financial sector services as well as purely lending and financing operations (IADB 2017). Included in the IADB's SDGs focus is the Independent Consultation and Investigation Mechanism, which exists to set internal lending policies and guidelines for the IADB as well as to monitor its compliance with these guidelines and to evaluate complaints brought by outside groups regarding the decision-making used by the IADB and the way in which financing is implemented throughout the lifecycle of the project.

Recently, the Asian Development Bank created what it defines as a technical assistance mechanism referred to as Supporting the Implementation of Environment-Related SDGs in Asia and the Pacific, in order to provide assistance on far more than a financial level (Asian Development Bank 2017). Through this mechanism, the Asian Development Bank functions as a supplier of information and assistance to regional member states at the same time that it acts in a global governance capacity to ensure that these member states are honouring their financial and other commitments.

In a similar vein, the IADB has created the NDC (nationally determined contributions) Invest system to assist member states in aligning their policies and practices with commitments made under the Paris Agreement and in the course of supporting the SDGs (IADB 2017). This program works at all levels of the project development, financing, and implementation cycle as well as at the level of providing assistance to particular state sectors and industries as they attempt to implement their commitments.

Similar to the World Bank Group, the Asian Development Bank has become involved in joint efforts at SDG implementation and governance throughout the region, making it at the same time a more pivotal actor and also subject to greater global governance systems itself as well as functioning as a global governance mechanism in itself (Asian Development Bank 2017). The IADB has become similarly engaged with regional and international groups which seek to address some or all of the issues contained in the SDGs and designated as of interest to the region (IADB 2017). This subjects the IADB to the oversight and governance mechanisms established in these groups and operations, casting further focus on its operations and requiring the IADB to act in compliance with the terms of other global governance mechanisms while still functioning as a global governance mechanism itself.

Environmental mechanisms

United Nations Environment

By its very nature, United Nations Environment – formerly the United Nations Environment Programme – has a direct link to and with the SDGs and thus the ability to profoundly impact on and be impacted by them. As part of the UN structure, UN Environment has a different relationship to the SDGs than other entities, as it is part of the same overall global governance mechanism for the SDGs. This is reflected in the specific indicators and aspects of the SDGs that have been assigned to UN Environment for implementation and monitoring (UN Environment 2016).

Despite this, it is essential to remember that UN Environment, like the United Nations Development Programme discussed below, was established well before the SDGs – or even the MDGs – were conceived of and had differing functions until 2015. Within the confines of the current framework used to assess the achievement and progress of the SDGs and associated indicators, UN Environment is required to oversee a number of topics and to create or utilize existing entities in its structure in fulfilment of this requirement (UN Environment 2016).

In addition to these implementation functions, UN Environment has become a bridging entity between states, global governance mechanisms, and even private funding sources in order to promote the accomplishment of the SDGs (UN Environment 2015). This includes promoting adaptation measures by states across all phases of development, providing guidance in and oversight of efforts by states to implement measures for climate resilience across a number of sectors, encouraging public and private investment in sustainable and green initiatives and resource use, and the implementation of reducing emissions from deforestation and forest degradation (REDD/REDD+) measures by states (UN Environment 2015).

Additionally, UN Environment works to promote and oversee risk reduction programs across a large segment of states and programs to define and promote response systems for instances in which natural and other disasters occur (UN Environment 2015). In each of these capacities, UN Environment acts as a collaborator and as a global governance mechanism, creating a duality that has been reinforced by the SDGs and that emphasizes the ways in which it has been changed as a global governance mechanism by the SDGs per se.

United Nations Development Programme

As with UN Environment, the United Nations Development Programme (UNDP) exists and functions within the ambit of the larger UN system that also oversees the SDGs. While this somewhat changes the inherent power dynamics from the international organization's perspective, it does not detract from the global governance function assigned to and carried out by UNDP.

Given that the UNDP's function is the UN's development branch, it is perhaps not surprising that the SDGs have had a significant impact on the ways in which it operates in terms of policy generation, advice giving, and global governance. In this context, it is also perhaps not surprising that the UNDP's role in the SDG structure involves the incorporation of environment and sustainable development principles and concerns with other essential areas of concern, such as human rights and conflict management/post-conflict settings (UNDP 2017).

As a result of the expanded role for the UNDP as a global governance mechanism, it has begun to create a more streamlined framework of offices and functions within states and at the larger bureaucratic level. This demonstrates one of the ways in which the SDGs have had a direct and palpable impact on the UNDP as a global governance mechanism generally and also in terms of its individual focus areas in particular (United Nations Development Programme 2017).

This is evident in the creation and application of the Integrated Results Framework to judge the achievement of the SDGs and associated targets/indicators throughout UNDP operations (UNDP 2017). It is further demonstrated in the creation and implementation of the Mainstreaming, Acceleration, and Policy Support system, which is intended to ensure that all aspects of the SDGs are implemented evenly across the UNDP's operations as a whole (UNDP 2017, 2018).

Implications and lessons for the future of global governance

The above has provided a discussion into how the SDGs and global governance interact and co-exist with each other. This discussion has demonstrated that there is necessarily a strong correlation between the two concepts and they are at certain levels symbiotic. At the same time, the above has highlighted how the SDGs have required global governance mechanisms, and international organizations in general, to alter their structures and focuses in order to align with the SDGs.

These understandings and discussions are of importance in themselves. Further, they are important because they demonstrate that when the international community creates a set of goals and objectives – even those that are inherently soft law and non-binding – it has the ability to change global governance in durable and fundamental ways. In this manner, systems such as the SDGs have impacts and the ability to shape policy and governance well beyond the designated lifespan given to them.

By understanding how and why these changes occur, scholars of global governance have the ability to analyse the SDGs and global governance mechanisms on a variety of levels and with a better sense of their temporal impacts. This, in turn, has the ability to generate more robust insights into how and why these impacts will persist as well as whether and how they will change in the future. It is, of course, impossible to predict the shape of the international community and the global governance mechanisms which will define it by 2030, let alone afterward. What is certain, however, is that the SDGs will leave an indelible and durable mark on concepts of global governance and the application of global governance mechanisms that extends well beyond 2030. This will define the future of global governance in general and will doubtless define the instruments that follow in the wake of the SDGs.

Acknowledgments

The author wishes to express her immense gratitude to the editors, Prof. John Ravenhill, and the entire BSIA community for the warmth and collegiality she has received. As the 2018–2019 Fulbright Canada Research Chair in Global Governance at the Balsillie School of International Affairs, the author is grateful for the opportunity to speak as a voice for the Fulbright Canada Research Chairs in Global Governance as they celebrate the BSIA at ten years old.

References

Asian Development Bank. 2017. *The Asian Development Bank's Transitional Results Framework, 2017–2020*. Manila: Asian Development Bank.

Asian Development Bank. 2018. "ADB and the Sustainable Development Goals". *Asia Development Bank*. Accessed 11 December 2018. www.adb.org/site/sdg/main.

High Level Political Forum on Sustainable Development. 2018. *Handbook for the Preparation of Voluntary National Reviews 2019*. New York: United Nations.

IADB (Inter-American Development Bank). 2017. *IDB Sustainability Report 2017; Delivering Climate Agenda for LAC: IDB Actions for 2020.* Washington DC: Inter-American Development Bank.

UN Environment. 2015. *Annual Report 2015.* Nairobi: UN Environment.

UN Environment. 2016. *SDG Indicators: UN Environment as Custodian Agency.* Nairobi: UN Environment.

United Nations. 2018. "Division for Sustainable Development Goals". *Sustainable Development Goals Knowledge Platform.* Accessed 11 December 2018. https://sustainabledevelopment.un.org/about.

UNDP (United Nations Development Programme). 2017. *UNDP: Key Partner in the Implementation of the SDGs.* Washington, DC: United Nations Development Programme.

UNDP (United Nations Development Programme). 2018. *UNDP: Financing the 2030 Agenda, An Introductory Guidebook for UNDP Country Offices.* Washington, DC: United Nations Development Programme.

UNGA (United Nations General Assembly). 2000. United Nations Millennium Declaration. A/RES/55/2, 18 September.

UNGA (United Nations General Assembly). 2015. *Transforming our World: The 2030 Agenda for Sustainable Development.* A/RES/70/1, 21 October. Accessed 11 December 2018. https://sustainabledevelopment.un.org/post2015/transformingourworld.

World Bank Group. 2018. *Implementing the 2030 Agenda: 2018 Update.* Washington, DC. World Bank Group.

WTO (World Trade Organization). 2015. "The WTO and the Millennium Development Goals" *World Trade Organization.* Accessed 11 December 2018. www.wto.org/english/thewto_e/coher_e/mdg_e/mdg_e.htm.

WTO (World Trade Organization). 2017. *Mainstreaming Trade to attain the Sustainable Development Goals.* Geneva: World Trade Organization.

WTO (World Trade Organization). 2018a. "Aid for Trade". *World Trade Organization.* Accessed 11 December 2018. www.wto.org/english/tratop_e/devel_e/a4t_e/aid4trade_e.htm.

WTO (World Trade Organization). 2018b. "The WTO and the Sustainable Development Goals". *World Trade Organization.* Accessed 11 December 2018. www.wto.org/english/thewto_e/coher_e/sdgs_e/sdgs_e.htm.

Index

Page numbers in **bold** denote tables, those in *italics* denote figures.

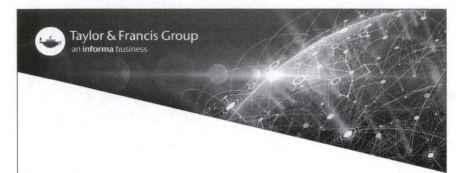